In Beaver World

BEAVER WORLD

In Beaver World

By
Enos A. Mills

With Illustrations from Photographs
by the Author

Introduction and Notes
by James H. Pickering

University of Nebraska Press
Lincoln and London

Introduction and notes copyright © 1990 by the University of Nebraska Press
All rights reserved
Manufactured in the United States of America

First Bison Book printing: 1990
Most recent printing indicated by the last digit below:
10   9   8   7   6   5   4   3   2   1

Library of Congress Cataloging-in-Publication Data
Mills, Enos Abijah, 1870–1922.
In beaver world / by Enos A. Mills; with illustrations from photographs by the author: introduction and notes by James H. Pickering.
p. cm.
Reprint, with new introd. Originally published: Boston: Houghton Mifflin Co., 1913.
Includes bibliographical references.
ISBN 0-8032-8172-2 (pbk.)
1. Beavers. 2. Mills, Enos Abijah, 1870–1922. I. Title.
QL737.R632M55   1991
599.32′32—dc20
90-35730   CIP

Originally published in 1913 by Houghton Mifflin Company

Seven photos appearing on unnumbered pages have been dropped from this Bison Book edition. Two photos have been added to the front matter.

∞

To

J. Horace McFarland

Contents

Introduction	xi
Working like a Beaver	1
Our Friend the Beaver	17
The Beaver Past and Present	37
As Others See Him	51
The Beaver Dam	63
Harvest Time with Beavers	81
Transportation Facilities	99
The Primitive House	117
The Beaver's Engineering	137
The Ruined Colony	151
Beaver Pioneers	173
The Colony in Winter	195
The Original Conservationist	211
Bibliographical Note	223
Notes	225
Index	231

Illustrations

Beaver World	Frontispiece
Enos Abijah Mills	xii
Longs Peak Inn in 1905	xvi
A Young Beaver on the side of a Beaver House	6
A Young Beaver Sunning Himself	22
In the Harvest-Field	32
Aspens cut by Beaver.	
A New Dam	66
Lake-Bed Canals at Lily Lake, October, 1911	102
Section of a 750-foot Canal at Lily Lake	102
Plan of Beaver Colony on Jefferson River, near Three Forks, Montana	108
An Unplastered and a Plastered House	124
The 334-foot Canal	140
Plan of Moraine Colony, with Dead-Wood Dam	144
The Dead-Wood Dam	148
House in Lily Lake	180

Introduction
By James H. Pickering

> I have never been able to decide which I love best, birds or trees, but as these are really comrades it does not matter, for they can take first place together. But when it comes to second place in my affection for wild things, this, I am sure, is filled by the beaver.
> —Enos A. Mills (1908)[1]

Second place or not, watching beavers was one of the life-long preoccupations of Enos Mills (1870–1922). From the time of his arrival in Longs Peak Valley in 1884 until the time of his death some thirty-eight years later, Mills kept year-round vigil on the ponds that lay nearby. No naturalist ever lavished more continuing attention on a single species of animal than Enos Mills lavished on the beaver. Though he often observed his quarry alone—for as many as sixty-four days in succession he tells us in the preface to *In Beaver World*—Mills took particular pleasure in introducing the beaver to others. Summer visitors to Mills's famous Longs Peak Inn could count on personally guided walks to the beaver ponds: to the Moraine Colony on Roaring Fork, some three and a half miles distant by way of the Longs Peak trail; to the colony at Lily Lake at the head of the valley; and to those closer by along Cow Creek (since 1961 Tahosa Creek) and Cabin Creek.[2] And, to make the experience even more memorable, visitors to the lodges on Cabin Creek were also conducted past "fire smutted stones" and the ruins of an old log cabin. Attached to two upright sticks was a wooden sign bearing the words "Kit Carson Cabin Site," intimating that the place had once served as home for the famous trapper, guide, and mountain man.[3] Enos Mills encouraged (perhaps even originated) the story that the legendary Car-

Enos Abijah Mills. From the *National Cyclopaedia of American Biography*, XX (New York: James T. White Company, 1929).

son had once trapped for beaver in Longs Peak Valley, for he found the subject of the beaver—past as well as present—an ever-fascinating one. He wrote about beavers regularly over the years, and by 1913 had a sufficient number of essays at hand to make them the subject of his third book, *In Beaver World*, published at Boston by Houghton Mifflin.[4]

Enos Mills's knowledge of beavers—like virtually everything else he came to know—was the result of self-education. Schooling of the formal sort had ended by the time Mills was fourteen, the victim of the same digestive ailment that disrupted his Kansas childhood and left him next to useless on the family farm near Pleasanton. That year, 1884, apparently as something of a last resort, Mills's parents allowed him to travel alone by way of Kansas City and Denver to the remote mountain valley nine miles south of Estes Park that he would one day introduce to the world. Though Mills would later exaggerate his degree of isolation and self-sufficiency—his father's cousin, the Reverend Elkanah J. Lamb (1832–1915), had been living with his family in Longs Peak Valley since 1875—the truth of the matter is that Mills landed in Colorado pretty much on his own, to do the best he could on a regimen of mountain air and spartan living.[5]

The years that followed were exhilarating ones of exploration and growth, culminating, fortuitously, in the discovery of vocation. From 1884 onward, the amphitheater-valley beneath the East Face of Longs Peak served as Enos Mills's home. It was a truly magnificent spot. To the west, towering nearly 5,000 feet from the valley floor and flanked by its near neighbors Mount Meeker and Mount Lady Washington, stood Longs Peak, at 14,255 feet the highest point in northern Colorado and arguably the single most beautiful mountain in a state filled with them. Across the valley to the east were the double heads and crests of the mountain called the Twin Sisters. A mile to the north, at the valley's entrance, was the hydrographic divide known as Lamb's Notch, separating the waters of the St. Vrain and Big Thompson rivers. To the south, the declension of an ancient moraine swept downward toward the heavily forested wilderness of Wild Basin, the watershed of the North St. Vrain River. Here amidst a world of wild flowers, quaking aspens, heavily wooded slopes of lodgepole pines, growths of juniper, green and shiny little carpets of kinnikinnik, and stands of graceful willows were to be found bighorn sheep, long-eared mule deer, together with the black bear, mountain lion, bob cat, wolf, fox, coyote, marten, and, of course, ponds of active beaver. Enos Mills carefully and systematically studied the faces and the ways of this mountain world, and over the years made it finally and uniquely his own. It was this world that he sought to share with others—through his public lectures and addresses, published articles and books, and the strenuous crusading efforts that led in 1915 to the creation of Rocky Mountain National Park.

In the mid-1880s, however, Enos Mills's major problem was a more prosaic one: how to gain sufficient health and make a living. It is not at all clear just how Mills achieved the former (though diet and exercise were apparently involved). What is clear is that whatever ailed the boy from Kansas was cured by the mountains, to the point that Enos Mills became capable of mountain-top exploits that seemed little short of miraculous. His claim to have climbed Longs Peak forty times alone and two hundred and fifty-seven times while guiding others, at every hour and in every season, and as many as three times in a single day, seems well founded, as do recitals of most of the other strenuous, often perilous, experiences that so impressed his reading and listening audiences. Employment was another matter. Estes Park shut down once summer visitors departed, and Mills was forced to extend his search for a livelihood to the copper mines of Butte, Montana. The Anaconda Copper Company offered what Mills wanted and needed most: winter employment at good wages, a flexible schedule that left summers free for the out-of-doors, the chance for career advancement, and, as an added benefit, access to a first-rate town library.

Books were important to Enos Mills from an early age, an inheritance, he said, from his mother. He read widely, if eclectically, and he read well: Shakespeare, Cervantes, Dickens, Thackeray, Macaulay, Byron, Keats, Shelley, Burns, Stevenson, Scott, Emerson, Irving, Whitman. Especially important were the nature writings of Henry David Thoreau, John Burroughs, and, particularly after 1889 and their chance encounter on a beach near San Francisco, John Muir. Though Mills's own books and articles give the impression of having been written almost exclusively out of the stuff of his own first-hand experience, the appearance is misleading. As *In Beaver World* with its concluding "Bibliographic Note" suggests, virtually everything Mills wrote was informed by his reading, though in most cases the exact authors and works consulted must be largely inferred. His personal library was large and impressive enough ("lined with a thousand books" one visitor reported) to do credit to any college professor. As a nature writer, as with everything he undertook, Enos Mills unquestionably was a quick study. But no one can trace the subjects and arguments of his articles and books, or chart the development of his highly readable prose style, without realizing that Enos Mills was a writer with clear literary antecedents.

Until he was almost twenty, Enos Mills's interest in nature, though greatly expanded by his travels and explorations, remained unfocused and undirected. His meeting with John Muir (1838–1914) in December 1889 changed that. Mills had gone to San Francisco apparently intent on enrolling in business college in order to improve his career advancement opportunities back in Butte. In California, he recalled, "on the beach near the old cliff house [in Golden Gate Park] I came upon a number of people around a small gray bearded little man who had a hand full of plants which he was explaining. . . . As soon as the people scattered I asked him concerning a long-rooted plant that someone had dug from a sand dune." Muir, already celebrated for his efforts to preserve the grandeur of Yosemite Valley, then proceeded to take Mills on a four-mile walk through the park to the end of the car line: "He said to me, 'I want you to help me to do something for parks, forests and wild life.'" Though Mills would later exaggerate the strength of their relationship, particularly during the years after 1894 when Muir's books began to appear, there can be no question that their moment together on a wintry beach at San Francisco gave purpose and direction to the whole of Enos Mills's subsequent life.[6] In its aftermath came two decades of purposeful activity that transformed an unknown young man from Estes Park into a public figure with a state and national reputation.

Longs Peak Inn in 1905. Courtesy of the National Park Service.

Mills began his literary career during the mid-1890s by publishing brief articles about Estes Park and Colorado scenery, illustrated with his own photographs, in the Denver papers. About the same time he started reporting the society news of Denverites summering in Estes Park, for which he was paid a third of a cent a word. Mills collected these news items by riding among the widely scattered resorts of Estes Park—to Elkanah Lamb's Longs Peak House, to the Elkhorn Lodge near Estes Park village, to Horace Ferguson's Highlands Hotel west of Marys Lake, to the Earl of Dunraven's English Hotel on lower Fish Creek road, and to Abner Sprague's ranch in Moraine Park. In the process Mills also began to pick up the bits and pieces of information about the early days of Estes Park that he would bring together in 1905 in his first published book, *The Story of Estes Park and a Guide Book*.[7] Before the decade was out Mills had launched his career as a writer that would result first in essays and articles for such periodicals as *Saturday Evening Post*, *Atlantic*, *World's Work*, *McClure's Collier's*, *Harper's*, *Country Gentleman*, *Sunset*, *Country Life*, *American Boy*, and *Youth's Companion* and then in a series of sixteen books published between 1909 and 1931 by Houghton Mifflin and Doubleday, Page and Company.[8]

Until 1902, Mills regularly inhabited the small tin-roofed homestead cabin built by his own hands in 1885–86. That year, as the culmination of what was obviously a lifetime dream, Enos Mills purchased from the Lambs the small resort hotel in Longs Peak Valley known as Longs Peak House.[9] Changing its name, Mills proceeded to enlarge the place, and despite a fire that destroyed the main lodge and its recently completed fifty-foot dining room in 1906, soon turned Longs Peak Inn into one of the most unique and distinctive hostelries in the nation. Visitors never ceased being impressed by what they found. Anna M. Rudy's account of her 1915 visit is typical:

> We rode from the village of Estes Park to Longs Peak Inn in a comfortable auto through one of the most picturesque portions of the Rocky Mountain National Park, arriving at five o'clock. I had time before dinner to revel in the charms of the Inn and its surroundings. It is the largest of a number of log houses placed

around on the slope in irregular order. It has quaint balconies on the greater part of two sides where one finds comfortable rustic chairs and writing tables. Not far away are groves of lodge pole pines and aspen trees, which Mr. Mills aptly calls "bare legged children," and all around are beautiful mountain flowers of many varieties.

Mr. Mills is his very own architect as well as superintendent of all the building, and does much of the work himself. In the Inn and his cabin, where he is at home all the year, you see exemplified William Morris' creed, "Have about you only the things that are useful or that you believe to be beautiful." In the living room of the Inn— a room approximately 30 x 40 feet in dimensions, all the furniture is made of wind or fire killed trees, with that soft indescribable gray tone that no skill of brush and palette could produce.[10]

The energetic Enos Mills, with his "ruddy face and a very broad forehead surmounted by a veritable shock of carrot colored hair,"[11] took personal charge of the welfare of his guests. The aim was to provide education through recreation, with not a minute wasted. In the early years of Longs Peak Inn, Mills made himself personally available for trips to the top of the peak (for a fee of twenty-five dollars), though later he turned the task of mountain guiding over to others. For those less strenuously inclined, there were day hikes to the nearby lakes, glaciers and lower elevations, as well as horseback rides, fishing, tennis, nature walks, and, of course, trips to visit the beaver ponds. In the evenings, after dinner, came Mills's famous fireside talks. Enos Mills was a genial and attentive host, particularly toward those who were willing to abandon themselves to the ways of the mountains.

For all his obvious success as a mountain inn-keeper, however, it seems likely that Mills's reputation would have remained little more than a local one had it not been for two critical events during the first decade of the twentieth century that catapulted him to state, regional, and national attention. The first of these was his appointment in 1902 as Colorado's official State Snow Observer, a career Mills followed for three successive seasons. The position of Snow Observer was a highly romantic calling, one that quickly captured the public imagination. And Enos Mills, always alert to opportunities for self-promotion, made the most of it. "The work of state snow observer," he told the *Denver Republican* in 1905, "made it necessary for me to ramble wild heights and to go beyond the trails in all kinds of weather. I traversed the forests, invaded the gulches, walked the bleak heights, scaled wintry peaks, went beyond the trails and visited the silent places. The duties of the snow observer are mostly confined to measuring snow accumulations at the head waters of streams."[12]

The reports that Mills sent on to Denver and the office of State Engineer L. G. Carpenter, the head of Colorado's Irrigation Department, indicate that he was well worth his hire. In his report of February 20, 1904, Mills informed Carpenter:

> I went from home S. W. four miles and up to an altitude of 10,000:—thence from home up the S. Poudre to Chambers Lake: then S. W. to a point on the N. W. slope of Hague's peak about 13,000 feet: S. W. to ditch camp—Grand—about 10,500 ft., Southward to summit of Specimen mountain then turning west near Poudre lakes to North form of Grand-Pacaiic slope—then timberline on Lead mountain, Richthofen and Lulu pass. From this pass 22 miles southward down N. Fork to Grand Lake:—four miles N. E. to Big Meadows then S. W. to Lehman, Coulter, then S. E. to Berthoud pass and Empire. Total traveled between home and Empire about 170 miles; 70 miles of this was on snow shoes.[13]

What follows are five pages of details reporting the average snow depth and wind, weather, and timber conditions. "12 inches of snow fell on me during the course of this trip . . . ," Mills concluded his report. "The coldest record while I was out was the night at Grand Lake when the thermometer descended to 28º below zero."

The "Snow Man" someone called him, and the epithet not only stuck but spread. What made Mills's winter trekking expeditions all the more remarkable was the apparent ease and simplicity of the undertaking. "I dress lightly," he wrote about 1905:

> Medium weight woolen flannels, canvas coat and overalls, German socks, high-cut overshoes and a slouch hat. I carry a sweater and an extra pair of overalls for emergencies. I rarely take anything but raisins for food. A pound of these will sustain me for a week. I al-

ways take two packages of matches and a compass. All bedding is
left behind, and firearms, I never carry. A small ax is ever with me,
and generally there is a candle or two in my pocket. When I have to
start a fire with damp wood in the midst of a raging blizzard, can-
dles are of inestimable value for kindling.[14]

Translated into the pages of the popular press the colorful adventures of "Colorado's Snow Observer" were well calculated to capture the reader's imagination.

By the time he retired from the post of "Snow Observer" in the
spring of 1906, the Enos Mills legend had been well and permanently launched. He was Enos Mills of Colorado: an intrepid, solitary, wilderness explorer of heroic dimensions, the survivor of
storm, avalanche, and a dozen death-defying mountain-top adventures.

The second important event of the decade extended Mills's reputation to the nation at large. Under the Theodore Roosevelt ad-
ministration, conservation and the creation of new forest pre-
serves had become a major part of the national agenda. Most of
these new national forests were, of course, to be found in the
West. There Roosevelt and Gifford Pinchot, the new head of the
Forest Service, not surprisingly encountered opposition among
established timber, mining, cattle, and water interests, who were
indignant at the thought of any curtailment of their freedom. To
counterbalance their arguments, a western spokesman for forest
conservation was needed, and Enos Mills of Colorado became
Roosevelt's and Pinchot's choice.

As an independent, salaried lecturer on forestry, a position he occupied from January 1907 to May 1909, Enos Mills literally
crossed the nation to promote the cause of forest conservation
and the recreational and spiritual uses of the wilderness. The
pace he set was frenetic. A surviving itinerary for the eight-month
period October 1908 to May 1909 shows well over a hundred
speaking engagements in more than thirty states plus the District
of Columbia, before women's clubs, school and college groups,
civic and business associations, and lecture societies (including
New York's Carnegie Hall).

Mills's most frequently announced topic was "Our Friends the Trees." His central message—the need to save and preserve America's endangered forests—seldom varied. It was a message that Mills, clad in brown sack suit and looking every bit the part, delivered in a manner that freely mixed fact and poetry. A reporter for the *Omaha World-Herald* in April of 1907 noted, "He made the subject as interesting as a romance, as instructive as an Emerson essay, as absorbing as a newspaper scandal."[15]

Mills took to the lecture platform with an enthusiasm that virtually insured his success. Yet, for all his growing public exposure and notoriety, and the apparent ease in which he entered the public arena, and later into public debate and controversy, Enos Mills was and remained to the end of his life a fundamentally shy and private man, content to hide (or take refuge) behind his own public persona and the colorful and romantic characterizations given him by others. As one would-be interviewer of the period 1911–1912 noted with obvious disappointment of his encounter with Mills, "He is long on enthusiasm, but short on autobiography."[16]

Mills left government employment in May 1909, only to take up a new challenge: the six-year campaign for the creation of a new national park in the Estes Park region—a task he later came to refer to as "the most strenuous and growth-compelling occupation I have ever followed"[17] and "the achievement of my life."[18] These efforts, which climaxed in the bill creating the Rocky Mountain National Park signed into law by President Woodrow Wilson on January 26, 1915, once again called on Mills's considerable public relations skills as a wilderness advocate. He tirelessly and totally committed himself to the undertaking. Before he was through, Enos Mills had made more than three hundred lecture appearances; written scores of newspaper and magazine articles, editorials, and letters to the editor; organized meetings and circulated petitions; and attempted to enlist the personal support of anyone of influence willing to listen. These efforts took Mills once again back and forth across the country, including Washington where, in a December 1914 appearance before the House Committee on Public Lands, he was accorded star billing as "one of the noted naturalists, travelers, authors, and lecturers of this country."[19] Victory did not come easily. Arrayed against Mills and other park supporters were a variety of agricultural and commercial interests that felt threatened by the prospect of yet another federal forest preserve.

It finally took three separate bills, five major revisions, and almost two full years to move the park legislation through the Congress. In the aftermath, however, it was clear enough to whom major credit belonged. On January 20, 1915, the *Denver Post*, in announcing the passage of the final bill two days before, carried on its front page a cartoon showing a smiling goddess labeled Colorado shaking hands with a bare-headed Enos Mills and saying, "Enos, I'm proud of you!" On September 4, 1915, on the occasion of the formal dedication ceremonies in Horseshoe Park at which he presided, the *Post* bestowed upon him the title "Father of Rocky Mountain National Park."

The final seven years of Enos Mills's life should have been ones of happiness and contentment. Though Mills finally married in August 1918, at the age of forty-eight, and the next year became a father,[20] these years were marred by events of controversy and acrimony with Mills himself at the center. Most Americans knew him as a generous and affable guide to the Colorado wilderness, but there was another, darker side to Enos Mills's personality. For all his many virtues, Mills was a highly temperamental man, quick to take offense, and all too often unwilling or unable to acknowledge the correctness of any views but his own. The truth of the matter is that Enos Mills was essentially a man of solitude, more at home alone on mountain heights than in the give and take world of men where it was necessary to deal with those of different, often opposing, ideas and interests. Mills did best when he was squarely in charge of things—as host at Longs Peak Inn, on the lecture platform, or in his writing. When he was not in control of a situation or event, he often showed himself remarkably thin-skinned and contentious. Quarrels with family and friends punctuated his early years in Estes Park.[21] Later, during the long, protracted campaign for Rocky Mountain National Park, he became openly critical of the Forest Service, whose "wise use" approach to forestry, Mills believed, placed it in direct opposition to his proposed park. "Scratch any old Forest Service man," he wrote in March 1911, "and you will find a Tartar who is opposed to all National

Parks."[22] Enos Mills had, in the words of James Grafton Rogers (1883–1971), the founding president of the Colorado Mountain Club, "a genius for making enemies," and, Rogers might have added, a genius for making them out of one-time friends and supporters.[23]

In the years after 1915, Mills's chief antagonist, ironically enough, was none other than the National Park Service, the agency which seemed most likely to effect and extend Mills's own wilderness agenda. The major cause of Mills's anger was the decision by Superintendent L. C. Way in the spring of 1919 to grant an exclusive transportation franchise agreement for travel within the Park. What outraged Mills was that Way's contract with the Rocky Mountain Parks Transportation Company excluded not only independent rental car ("jitney") drivers but touring cars owned and operated by established hotel men like Mills himself. Mills sued Way and the Park Service in the U.S. District Court of Colorado for abrogating his "common rights as a citizen of the State of Colorado in traveling over the Park roads." Gradually the controversy assumed larger dimensions and became involved in a suit by the State of Colorado challenging the right of the federal government to regulate traffic over roads never formally placed under United States jurisdiction. By the time the so-called "Cede Jurisdiction" controversy worked its way to a close in February 1929, however, Enos Mills was dead.

Mills's death came suddenly and without warning on the evening of September 21, 1922. Though he had been ailing much of that summer, worn out from a new round of hectic cross-country travel to take to the people his case against the Park's transportation monopoly, the setback seemed only temporary. But, as summer turned to fall, Enos Mills, a man who had prided himself on his extraordinary stamina and good health, did not recover. Death came unexpectedly in the form of a heart attack, in all probability brought on by blood poisoning. Following a short, secular service in the main lobby of the Inn, Mills's body was carried across the road and buried next to his homestead cabin.[24] He was fifty-two.

Mills's maturation as a nature writer coincided almost exactly with the beginning of his lobbying efforts on behalf of the Rocky Mountain National Park. Though he had previously enjoyed good success in selling his essays to popular magazines, it was not until 1909 that Mills was able to persuade a major publisher to issue a collection of his work. Mills chose his publisher well. Boston's Houghton Mifflin Company, which included among its authors "the two Johnnies," Muir and Burroughs, was America's foremost nature publisher. That Mills's first books bore the Houghton Mifflin imprint virtually assured that they would be noticed.

Wild Life on the Rockies (1909) and its sequel *The Spell of the Rockies* (1911) established a clear pattern for the books that followed. Enos Mills was a consummate teller of stories, and, as his readers discovered, the Rocky Mountain world he told about was one that the author intimately knew and loved. The world to which Enos Mills introduced his readers was a strange and wonderful world apart—whose remote and quiet wilderness, as Mills makes clear, was being inexorably tamed and forever altered by the forces of twentieth-century life. What Mills offered was a most palatable introduction: a miscellany combining nature essays with exciting first-hand accounts of the author's own adventures among Colorado's peaks and forests. Above all, Mills's essays celebrated the magnetic, transforming bond or "spell" between man and nature, together with the aesthetic and spiritual values to be gained from becoming attuned to nature's ways.

What clearly appealed most to Mills's readers, however, was the character and personality of the author himself: his boyish, thoroughly engaged, enthusiasm; his never-ending sense of wonder and excitement; his abundance of goodwill and buoyant optimism; his relish for new experience; and his courage and undauntedness even in the tightest places. So attractive, in fact, was the figure that Mills placed before them that most readers and reviewers seemed scarcely aware of the author's literary and scientific shortcomings and deficiencies.[25]

In Beaver World (1913), however, is a different kind of book. In his two earlier books, Enos Mills the mountain adventurer is never long absent from the scene. Here, by contrast, Mills is content to focus the reader's attention on the subject at hand and to adopt the role of quiet observer. His purpose is to tell the story of the beaver rather than his own, and in telling that story Enos Mills produced what may well be his finest sustained piece of nature writing.

Part of the success of *In Beaver World* resulted from the fact that the subject of the American beaver, even in 1913, was still remarkably fresh ground. As the *Nation* correctly observed in its July 1913 review of Mills's book, "Except for a few good notes in magazine articles, chiefly valuable for their photographic illustrations, nothing of much account has been published about the beaver since the classic book by Lewis H. Morgan, issued in 1868."[26] Though Morgan (1818–1881), who like Mills lacked formal scientific training as a naturalist, humbly admitted in his Preface that *The American Beaver and His Works* "falls much below the dignity and completeness of a monograph,"[27] his book remained almost half a century later the single most exhaustive work on the subject.

Mills took his assignment seriously and did his work well. He read Morgan's book, and whatever else he could find on the subject, and these investigations he combined with his own twenty-seven years of observations and fieldwork. Considering the limitations under which he worked, the results are impressive. The basic factual information about the beaver that Mills provides— *e.g.*, their number of teeth, the months in which they are born, their body weight and length, their longevity, their tail morphology, the properties of their newborns, their diving times, etc.— seem entirely accurate when measured against other, more recent, accounts.[28] There are, predictably, some errors of fact. Mills is wrong, for example, when he asserts (p. 40) that the South American beaver "may be called a link between the muskrat and the beaver." At best, however, this is an esoteric point. We cannot, of course, check the accuracy of his descriptions of individual beavers, colonies, dams, channels, or behaviors. But what we *can* check suggests that Enos Mills was very much committed to scientific accuracy in his descriptive work.[29]

To be sure, there are problems. From the perspective of the modern naturalist, Mills can be faulted for an overreliance on unique and unsynthesized events as the source for his understanding and ideas. Much of *In Beaver World* consists of descriptions of this and that dam, this and that colony, this and that beaver, etc., and in their examination Mills rarely takes a statistical or experimental (as opposed to a descriptive) approach. There are, however, some notable exceptions; and these exceptions go a long way toward establishing that Mills was capable of making sound scientific judgments. For example, Mills asks in the chapter titled "The Beaver Past and Present" (p. 44ff) whether beaver are weather-wise and answers in a most modern manner by reporting on the variation in behavior among beaver colonies that are all exposed to the same weather. He marks trees by notching them in "Harvest Time with Beavers" in order to determine how far beavers might transport logs, and then searches for notched logs (p. 85ff). And in "The Primitive House" he takes a statistical approach to the question of how commonly beaver houses are plastered in the fall (p. 124).

The modern scientist would also fault Mills because he has a ready, adaptive, and often teleological explanation for each observation. Rarely does Enos Mills leave questions open. One of the few exceptions occurs on page 115 ("Transportation Facilities) where he asks whether a tunnel increased the water supply of a colony by design or by accident. At this point in his discussion Mills does seem to realize that he has framed the debate between the teleological view and the behaviorist view, but here the matter ends. Usually, however, his mind seems uncluttered by such concerns.

The controversial center of *In Beaver World*, as Enos Mills himself was clearly aware, is the author's willingness to grant his subject the ability to reason. On page 57 he states his position succinctly in a single paragraph in the chapter titled "As Others See Him":

> For more than a quarter of a century I have been a friendly visitor to his colonies, in which I have lingered long and lovingly. That he makes mistakes is certain, but that he is an intelligent, reasoning animal I have long firmly believed. As I said in *Wild Life on the Rockies*,—"I have so often seen him change his plans so wisely and meet emergencies so promptly and well that I can think of him only as a reasoner."

Here Enos Mills the naturalist was walking on most dangerous ground. Specifically, he risked becoming directly embroiled in the so-called "nature-faker" controversy begun by John Burroughs in March 1903, when he used the occasion of an article in the *Atlantic Monthly* to ridicule as "sham naturalists" those who attempted to humanize and sentimentalize animals by giving them human traits. Burroughs's essay, and the spirited, at times rancorous, debate encouraged by its inevitable sequels and rebuttals (including the famous sortie into the field made by Theodore Roosevelt), engaged the attention of the popular press at precisely the time that Enos Mills was attempting to establish his own authority as a naturalist.[30]

When approached for his views, Mills initially tried to straddle the issue, perhaps because of his own genuine ambivalence.[31] Nevertheless, his early writings betray the tendency toward the anthropomophic treatment of wild life. On a number of occasions in *Wild Life on the Rockies,* Mills was willing, if perhaps at times unconsciously, to attribute the ability to think not only to beavers but to other animals as well—a willingness which moved one reviewer to comment that such "utterances, of course, put Mr. Mills in the 'reason' school of American nature writers, as opposed to the 'instinct' school of which John Burroughs, the Sage of Slabsides, is the official mouthpiece."[32]

It is unlikely that Mills arrived at his anthropomorphizing through any careful, systematic weighing of abstract, theoretical arguments for or against positing mental states for animals. More likely, in the beginning, he simply followed his "gut feelings," much in the way many modern pet owners do with respect to a favorite dog or cat, without any real in-depth understanding of the points of argument that could be raised against his position. Though many pages go by in *In Beaver World* without significant anthropomorphizing, it is unmistakably there. At times (particularly in the chapter "The Beaver's Engineering") Mills limits himself to the use of concepts like "intention" and "forethought," which at least some naturalists of Mills's own day would have been willing to defend. In this chapter, for example, Mills finds himself describing a protracted "project" carried out by a whole group of beavers in which many individual actions seem to be guided by a long-range goal. These behaviors, Mills came to believe, could not be explained simply as responses to short-term needs or stimuli. Rather, many of the actions made sense only in the context of "intention" and "forethought," of promoting rewards that would be received many months (even years) in the future. On other occasions, however, Mills's anthropomorphizing leaves entirely the field of defensible debate. He puts words into beavers's mouths (p. 4); terms beavers "friends" (p. 19); commends them as ethically or morally honorable (p. 30); and calls them "people" (pp. 98, 154) or "folks" (p. 87). The problem with wandering so "near the danger zone" in his treatment of animals was that Mills, who on other occasions took great delight in correcting popular misconceptions about the natural world, risked not only straining his readers' credibility but undermining the very scientific understanding of nature that he so wished to encourage.

Enos Mills was well aware that he was courting trouble. Just as soon as he has voiced his belief in the beaver as "reasoner" (in the chapter "As Others See Him"), Mills immediately turns for support to two highly regarded authorities on the subject of beavers and animal rationality, Lewis H. Morgan and George John Romanes.[33] Fortunately, however, Mills's anthropomorphic tendencies seem to have been largely overlooked by the contemporary reviewers of *In Beaver World*, who seemed most impressed by the fact that the book "is not dry science." As the *Charleston News and Courier* continued, "It is like everything else he has done, full of life and action and human interest, and it is at the same time authoritative."[34] When Mills's anthropomorphism was noted, in fact, it seems to have even evoked a positive response. In declaring the beaver "a reasoner," Chicago's *Inter Ocean* noted in a full-page review on June 29, 1913, that

> Mr. Mills flings down the gauntlet to Theodore Roosevelt, John Burroughs and others of the "instinct school" who hold that instinct alone rules the animal world and cry "nature fakers" at William J. Long, Ernest Thompson Seton, Charles D. Roberts and others of the "reason school" of naturalists. Mr. Mills writes with reserve and refrains from trying to read reason into all the works of the beaver. Nonetheless, the facts in his book will make the "instinct school" do some hard thinking.[35]

Introduction   xxix

The question of animal rationality is, of course, by no means a closed case in current scientific circles,[36] and Mills, interestingly enough, raises some of the very issues that biologists today find to be of interest. Moreover, it is pointless for anyone to quibble greatly with a work published more than three-quarters of a century ago that is, in so many ways, clearly a success. The time that Mills spent in the field carefully observing, photographing, taking measurements, and drawing diagrams of the beaver was time well spent, and his willingness to study these animals in their natural environment anticipates the development of ethology in the 1930s. Though some of his work has been superseded over time, the voice of Enos Mills remains an historically important one. Those who come this summer or next to the beaver ponds in the Hidden Valley area of Rocky Mountain National Park, deep in that part of the world that Enos Mills loved most, will find that the Park Service has inscribed sentences from *In Beaver World* on permanent markers to help educate visitors in the ways of that animal. As those signs attest, the wonderful and enduring strength of *In Beaver World* resides, finally, in the accounts that Enos Mills provides of beavers in their primordial state, accounts that not only continue to make for good and interesting reading but that retain considerable value for any modern day biologist doing work on this most fascinating wilderness creature.

Notes

1. "The Beaver and His Works," first published in the December 1908 issue of *The World To-Day*. Mills later made the essay part of his first published volume of collected nature essays, *Wild Life on the Rockies* (1909).

2. The beaver ponds that Mills celebrates in *In Beaver World* are described by Edward R. Warren in his "Notes on the Beaver Colonies in the Longs Peak Region of Estes Park, Colorado," *Roosevelt Wild Life Annals*, 1-2 (1926): 192-234. In the summer of 1922, Warren, a biologist attached to Syracuse University, studied in detail the beaver population along Cow Creek, Roaring Fork, Cabin Creek, and at Lily Lake; he concluded that there were some 160 beavers living in Longs Peak Valley, occupying 20 active lodges. Warren found Enos Mills a willing source of information, but "owing to ill health he was unable to go out into the field, though he expressed the hope and intention of making further studies

on beaver in the autumn after the Inn had closed for the season." That was in August; a month later, on September 21, 1922, Enos Mills was dead. Warren's view of Mills's works on the beaver (which he cites in his bibliography together with Lewis H. Morgan's classic 1868 study *The American Beaver and His Works*) is a complimentary and uncritical one.

3. The legend of Kit Carson's cabin and Enos Mills's role in its creation is discussed in the notes to my edition of Joe Mills, *A Mountain Boyhood* (Lincoln: University of Nebraska Press, 1988), pp. 294–95. The claim has little, if any, historical support.

4. With two exceptions—*In Beaver World* (1913) and *The Grizzly: Our Greatest Wild Animal* (1919)—Mills's books took the form of collected essays on a variety of subjects. The exceptions suggest the importance he attached to beavers and bears, respectively.

5. The single most accessible source of information on the life and career of Enos Mills is found in the Enos Mills Papers, a collection consisting of manuscripts, correspondence, speeches, biographical data, and clippings and articles by and about Enos Mills, mounted in a series of scrapbooks that Mills's widow, Esther Burnell Mills (1889–1946), presented to the Denver Public Library. It is now housed in the Library's Western History Department. Hereafter cited as the Enos Mills Papers.

6. Mills never elaborated on his meeting with Muir, though he referred to it repeatedly, if elliptically, throughout his career. The following brief undated autobiographical paragraph found among the Enos Mills Papers is typical:

> I suppose fairies led me to California and John Muir. He gave purpose to my bent and said to me, "I want you to do something for parks, forests and wild life." This was in 1889. Up to the time of his death he was an excellent coach in seeing that I did it. He insisted that I must learn to write and speak. He early became the factor in my life.

Nonetheless, the precise nature of the Muir-Mills relationship remains unclear. The surviving correspondence between the two (located in the Pacific Center for Western Historical Studies at the University of the Pacific in Stockton, California) is formal, brief, sporadic, and without substance or depth. It consists of thirteen pieces: the earliest a letter from Muir dated December 31, 1902, giving Mills permission "to quote as much as you like from my National Park [*Our National Parks*, published in 1901] for a review"; the final piece, a letter from Mills to Muir dated November 24, 1914, just prior to Muir's death. Only one exchange—which came during the winter of 1913—suggests the kind of mutual regard and personal warmth that Mills himself undoubtedly would have liked to have enjoyed. Mills told Muir in his letter of January 31, 1913, that "You have helped me more than all the others; but for you I might never have done anything for scenery." To which Muir responded on February 16, 1913: I always feel good when I look your way: for you are making good on a noble career. I glory in your success as a writer & lecturer & in saving God's park, for the welfare of humanity." Though he visited Muir in California on several occasions, Mills was never able to lure Muir to Estes Park.

7. The first recorded discovery of Estes Park (park means "valley" in the vernacular of the mountains) came in October 1859, when Joel Estes (1808–1875) and one of his sons, following the Little Thompson River westward into the mountains on a hunting and exploring expedition, entered the valley that would later bear his name. Estes liked the place and built several cabins, corrals, and outhouses on what is now lower Fish Creek road (the site is near the place where Highway 34 from Lyons enters the valley), and for a number of years ran cattle and took in tourists. The first major influx of permanent residents into Estes Park dates from the 1870s. The best single account of the early history of Estes Park is found in C. W. Buchholtz, *Rocky Mountain National Park: A History* (Boulder: Colorado Associated University Press, 1983).

8. Twelve of these books were published during Mills's lifetime. The final four volumes were published between 1923 and 1931 by Esther Burnell Mills, from materials that she brought together and edited.

9. Shortly after he arrived in Estes Park in 1875, the Reverend Elkanah J. Lamb opened his ranch, which he named Longs Peak House, to the public, and for a number of years supplemented his minister's salary (at five dollars a trip) by guiding parties of tourists to the top of the peak. "If they would not pay for spiritual guidance," he later noted in his memoirs, "I compelled them to divide for material elevation." Elkanah Lamb, *Memoirs of the Past and Thoughts of the Future* (United Brethren Publishing House, 1906), p. 166.

Thanks to its strategic location at the foot of Longs Peak, Lamb's ranch soon established a reputation as a good base of operations for those setting out for the summit, some eight miles distant. It was from Longs Peak House that the intrepid English adventurer Isabella Bird, accompanied by the mysterious and legendary character known as "Rocky Mountain Jim" Nugent, set out on her famous October 1873 conquest of Longs Peak (which she so graphically described in *A Lady's Life in the Rockies*, 1879). Father Lamb eventually turned the guiding over to his son, Carlyle Lamb (1862–1958), and it was Carlyle who in 1885 conducted the fifteen-year-old Enos Mills on his first trip to the top of Longs

Peak and two years later accompanied a twenty-five-year-old Connecticut druggist named Frederick Hastings Chapin (1852–1900) to the same destination. Chapin made this trip the focus of his *Mountaineering in Colorado: The Peaks about Estes Park*, the first classic of Colorado mountaineering. Originally published in 1889 by the Appalachian Mountain Club, Chapin's delightful book was reprinted, with introduction and notes by the present author, by the University of Nebraska Press in 1987.

It was Carlyle Lamb rather than his father who in 1889 had actually secured formal title to the 160-acre tract on which Longs Peak House stood. Enos Mills's own homestead cabin stood directly to the east and across the road. After Carlyle turned the Lamb ranch over to Mills in 1902, his parents built a new ranch called Mountain Home about a mile to the north at the head of Wind River, a place long known as Lamb's Notch (in more recent years the site of Wind River Ranch).

10. Anna M. Rudy, typescript manuscript, "Prepared for the Woman's Literary Club, Colorado Springs, 1915," Enos Mills Papers. For Mills's own description of the Inn, see Enos A. Mills, "A Home of Forest Fire Logs," *Sunset Magazine* (May 1921).

11. Enos Mills Papers.

12. Enos Mills Papers.

13. Enos Mills Papers. Mills moved quickly. Less than three weeks later, in a letter dated March 9, 1904, he reported to Carpenter that "I rambled through Sangre de Cristo without using either snow shoes or overshoes. . . . About 140 miles travel between Westcliff and Fort Garland."

14. Enos A. Mills, "Some Experiences of Colorado's Snow Observer," *Physical Culture* (c. 1905): 97. This five-page published account, illustrated by Mills's own photographs, is included in the Enos Mills Papers.

15. Unsigned review, Enos Mills Papers.

16. Unsigned review, November 9, 1907, *Atlanta Georgian and News*, Enos Mills Papers. Mills's public reticence and modesty did little to discourage stories about him. "People who go to Estes Park, in Colorado," reported Chicago's *Dial* on June 1, 1909, "hear the story of Mr. Enos Mills that when he was Government Snow Observer for that State Mr. Roosevelt telegraphed him 'Come at once to Washington,' and that he replied, 'Can't; I'm too busy.'" Enos Mills Papers.

17. Enos A. Mills, "Who's Who—and Why: Enos A. Mills Himself, By Himself," *Saturday Evening Post* (September 1, 1917): 9.

18. Undated autobiographical sketch, Enos Mills Papers.

19. Quoted in Patricia Fazio, "Cragged Crusade: The Fight for Rocky Mountain National Park, 1909–1915," master's thesis, University of Wyoming, Laramie, 1982, p. 135. For other accounts of Mills's role in the es-

Introduction   xxxiii

tablishment of Rocky Mountain Park, see Buchholtz, pp. 126–37, and Lloyd K. Musselman, *Rocky Mountain National Park: Administrative History, 1915–1965* (Washington, D. C.: U. S. Department of the Interior, National Park Service, 1971), pp. 17–27.

20. Mills married Esther Burnell (1889–1946) on August 12, 1918. A native of Kansas, she had come to Longs Peak Inn during the summer of 1916 to recover from a nervous breakdown, did some part-time secretarial work for Mills, and subsequently stayed on to take up her own homestead near Castle Mountain off Fall River Road. Their daughter, Enda, was born on April 27, 1919.

21. The story of Mills's quarrel with his uncle, the Reverend Elkanah Lamb, who had overseen his early days in Estes Park and accompanied him to Europe in 1900, is told in the introduction to my edition of *Wild Life on the Rockies* (Lincoln: University of Nebraska Press, 1988), pp. xxxviii-xl. The disrupted relationship between Enos and his younger brother, Enoch "Joe" Mills (1880–1935), is discussed in my introduction to Joe Mills's *A Mountain Boyhood* (1926; reprint, Lincoln: University of Nebraska Press, 1988).

22. Enos A. Mills to J. Horace McFarland, March 20, 1911. Papers of J. Horace McFarland, File 80, Division of Public Records, Pennsylvania Historical Museum Commission, Harrisburg, Pennsylvania. McFarland (1859–1948) was president of the American Civic Association, whose strong pro-conservationist stance made it a powerful ally in the campaign to establish the Rocky Mountain National Park.

Though the Forest Service, under both Pinchot and Henry Graves, who succeeded him as chief forester in February of 1910, professed an interest in national parks and scenic preservation, its main commitment was to a policy of "preservation through use" that made the nation's forest reserves available to lumbering and grazing interests under controlled, "scientific" conditions. Such a posture ultimately placed the Forest Service on a collision course with strict preservationists like Enos Mills.

23. The best example of Mills's ability to alienate totally one-time friends and supporters is provided by his relationship with J. Horace McFarland. Interestingly enough, the McFarland that Enos Mills excoriates so abrasively in letters of 1920 was the same man to whom he had dedicated *In Beaver World* seven years before.

24. A year later, Esther Mills, apparently fearing that vandals might desecrate the grave, had her husband's remains removed and cremated. The ashes were returned to Longs Peak and scattered. Esther Mills continued to operate Longs Peak Inn until 1945. A year later, the Inn burned to the ground.

25. Chief among these were Mills's tendency toward abstraction, didacticism, sentimentality, and clichés. His childlike optimism, however engaging, posed a problem as well, to the extent that it suggested a mind that refused to deal with the darker, harsher realities of life—to contemplate what Thoreau called "the other side of the mountain." Of his anthropomorphism more will be said below. Over the years Mills unquestionably became an abler, more interesting, and more confident writer, who not only curbed his early excesses but became much more precise, detailed, and objective in reporting the world of nature. For assessments of Mills's talents as writer and naturalist see Carl Abbott, "The Literary Career of Enos Mills," *Montana: The Magazine of Western History*, 31 (April 1981): 2–15; and Peter Wild, *Enos Mills* (Boise: Boise State University, 1979).

The question of Mills's ability as a natural scientist is critically important as well, and is only briefly touched on here—and then primarily in the context of his earliest books. Mills's preference for the "poetic interpretation of nature" tells only part of the story. The assessment of Mills's standing as a naturalist is complicated by the fact that, like Muir and Burroughs before him, Enos Mills was an all-purpose naturalist and as such something of an anachronism even in his own time.

26. Unsigned review, *Nation* (July 3, 1913), Enos Mills Papers. A survey of literature on the beaver, complete to its date of publication, can be found in Lee E. Yeager and Keith G. Hay, *A Contribution toward a Bibliography on the Beaver*, State of Colorado, Department of Game and Fish, Technical Bulletin, No. 1 (September 1955). The Enos Mills Papers also include reviews of *In Beaver World* from the *Denver News* (March 17, 1913), the *Chicago Inter-Ocean* (March 23, 1913, and June 21, 1913), the *Washington Evening Star* (March 29, 1913), the *San Francisco Chronicle* (March 30, 1913), the *Chicago Record-Herald* (April 3, 1913), the *New York Sun* (April 5, 1913), the *Peoria Journal* (April 12, 1913), the *Charleston News and Courier* (April 13, 1913), the *Boston Globe* (April 19, 1913), the *Boston Transcript* (April 21, 1913), the Chicago *Dial* (May 1, 1913), the *St. Louis Post-Dispatch* (May 3, 1913), the *Detroit News-Tribune* (May 4, 1913), the *New York Tribune* (May 10, 1913), the *New York Times* (May 18, 1913), the *Columbia* (s.c.) *State* (May 18, 1913), the *Des Moines Capital* (June 11, 1913), the Chicago *Continent* (June 12, 1913), *Outdoor Life* (June 1913), the *Indianapolis News* (August 2, 1913), the *Churchman* (August 23, 1913), the *Colorado Springs Evening Telegraph* (December 20, 1913), the London, England *Country Gentleman* (March 14, 1914), the *Literary Digest* (May 9, 1914), and the *Presbyterian Banner* (Pittsburgh, n.d.).

27. Lewis H. Morgan, *The American Beaver and His Works* (Philadelphia: J. B. Lippincott and Company, 1868), p. 6.

28. For my comments on the scientific accuracy and value of Mills's *In Beaver World* here and below I am greatly indebted to a former colleague and long-time friend, Dr. Richard Hill, professor of zoology and museum curator at Michigan State University in East Lansing. I am also indebted to Dr. Glenn Aumann, professor of biology and associate vice president at the University of Houston, for reading Mills's book and giving me his appraisal.

29. We do have one nearly contemporaneous study of the beaver colonies in the vicinity of Longs Peak that serves as a check on the validity of Mills's own. As indicated above in note 2, during the summer of 1922 Edward Warren, of the Roosevelt Wild Life Forest Experiment Station of Syracuse University, visited the same ponds that Mills described for the expressed purpose of making certain "that future students can make comparisons and note changes which have taken place." Warren's study of the beavers and their habitat in Longs Peak Valley is well documented by photographs, and, as might be expected, is far more scientifically detailed and descriptive than Enos Mills's book. Significantly, although Mills's observations are cited throughout, they are neither challenged nor contradicted by Warren.

30. John Burroughs, "Real and Sham Naturalists," *Atlantic Monthly*, 91 (March 1903): 298–309. The term "nature faker" was coined by Theodore Roosevelt in his essay "Nature Fakers," *Everybody's Magazine*, 17 (September 1907): 427–30. See also Paul Russell Cutright, "The Nature Faker Controversy," *Theodore Roosevelt the Naturalist* (New York: Harper and Brothers, 1956), pp. 126–39.

31. The heated discussion between President Roosevelt and the Reverend William J. Long (1866–1944), a Connecticut Congregational minister and author of a series of nature books including *Ways of Wood Folks* (1899), occupied much space in magazines. Long's tendency to attribute human habits and instincts to wild animals made him the lightning-rod figure of the "nature faker controversy." Mills was neutral, not wholly supporting either Roosevelt or Long.

32. Unsigned and undated review, *Inter-Ocean*, Enos Mills Papers. The reviewer continued: "And, worse than that, Mr. Mills's name has presumably already gone on the blacklist as a 'nature faker' of the most reckless sort, for some of the things he tells are not the kind that would be apt to happen at Slabsides." Slabsides was the name that John Burroughs had given to his rural New York retreat.

33. In a series of books written during the 1880s, beginning with *Animal Intelligence* (1882), George John Romanes (1848–1894) pioneered the new science of "comparative physiology," which attempted to provide accurate description of the mental states of animals on the basis of infer-

ences derived from their observed behavior. Though criticized for his methodology, rigor, and its resulting anthropomorphism, Romanes occupies an historically important place in the study of animal behavior. Interestingly enough, Mills apparently was not familiar with the corrective principle of "parsimony" developed by Romanes's younger colleague and subsequent critic C. Lloyd Morgan (1852–1936) which states, simply, that "In no case may we interpret an action as the outcome of the exercise of a higher psychical faculty, if it can be interpreted as the outcome of the exercise of one which stands lower in the psychological scale." See Philip H. Gray, "The Morgan-Romanes Controversy: A Contradiction in the History of Comparative Psychology," *Proceedings of the Montana Academy of Sciences*, 25 (1963): 225–30.

34. Unsigned Review, *Charleston* (s.c.) *News and Courier*, April, 13, 1913, Enos Mills Papers.

35. Unsigned review, *Chicago Inter-Ocean* (June 29, 1913), Enos Mills Papers. William J. Long (see note 31 above) together with Ernest Thompson Seton (1860–1946) and Charles D. Roberts (1860–1943) were the trio of nature writers that John Burroughs had singled out by name in his 1903 article as "sham naturalists" for their "biographies" of wild animals. Interestingly enough, both Seton and Roberts are cited in the "Bibliographic Note" that concludes Mills's *In Beaver World*. The works in question include Seton's *Wild Animals I Have Known* (1898), *The Trail of the Sandhill Stag* (1899), *Biography of a Grizzly* (1900), *Lives of the Hunted* (1901), and *Animal Heroes* (1905); and Roberts's *The Kindred of the Wild* (1902), *The Hunters of the Silences* (1907), *Babes of the Wild* (1912), *The Feet of the Furtive* (1913), and *Hoof and Claw* (1913).

36. For example, Donald R. Griffin, whose credentials as a modern behavioral scientist and a member of the National Academy of Science are impeccable, has urged his fellow scientists to open their minds to the possibility of mental states in nonhuman animals, thus breaking with many decades during which the behavioral sciences dogmatically insisted that it was highly inappropriate to posit consciousness, awareness, mental states, etc., in animals. Griffin writes:

> The possibility that animals have mental experiences is often dismissed as anthropomorphic because it is held to imply that other species have the same mental experiences a man might have under comparable circumstances. But this widespread view itself contains the questionable assumption that human mental experiences are the only kind that can conceivably exist. This belief that mental experiences are a unique attribute of a single species is not only unparsimonius; it is conceited. It seems more likely than not

Introduction   xxxvii

that mental experiences, like many other characters, are widespread, at least among multicellular animals, but differ greatly in nature and complexity. . . .

Opening our eyes to the theoretical possibility that animals have significant mental experiences is only a first step toward the more difficult procedure of investigating their actual nature and importance to the animals concerned. Great caution is necessary until adequate methods have been developed to gather independently verifiable data about the properties and significance of any mental experiences animals may prove to have.

Thanks to efforts like Griffin's, it seems safe to say that scientists are much more open than they were two decades ago to the *possibility* that mental states and awareness are real, causative agents in the behavior of some nonhuman animals. The very discussion, of course, goes far towards rescuing the scientific reputation of early naturalists like Enos Mills. See Donald R. Griffin, *The Question of Animal Awareness: Evolutionary Continuity of Mental Experience,* Revised and Enlarged Edition (New York: Rockefeller University Press, 1981), p. 170. See also Griffin's *Animal Thinking* (Cambridge: Harvard University Press, 1984) and Stephen Walker's, *Animal Thought* (London: Routledge and Kegan Paul, 1983). Walker concludes his study by observing that although "it is . . . reasonable to say that animals do not think as we do . . . , tacit mental organisation is evident not only in ourselves but in many other species. . . . Our organ of thought may be superior, and we may play it better, but it is surely vain to believe that other possessors of similar instruments leave them quite untouched" (pp. 387–88).

Preface

THIS book is the result of beaver studies which cover a period of twenty-seven years. During these years I have rambled through every State in the Union and visited Mexico, Canada, and Alaska. In the course of these rambles notice was taken of trees, birds, flowers, glaciers, and bears, and studious attention devoted to the beaver. No opportunity for beaver study was missed, and many a long journey was made for the purpose of investigating the conditions in live colonies or in making measurements in the ruins of old ones. These investigations were made during every season of the year, and often a week was spent in one colony. I have seen beaver at work scores of times, and on a few occasions dozens at one time.

Beaver have been my neighbors since I was a boy. At any time during the past twenty-five years I could go from my cabin on the slope of Long's Peak, Colorado, to a number of colonies

Preface

within fifteen minutes. Studies were carried on in these near-by colonies in spring, summer, autumn, and winter.

One autumn my entire time was spent in making observations and watching the activities of beaver in fourteen colonies. Sixty-four days in succession I visited these colonies, three of them twice daily. These daily investigations enabled me to see the preparations for winter from beginning to end. They also enabled me to understand details which with infrequent visits I could not have even discovered. During this autumn I saw two houses built and a number of old ones repaired and plastered. I also saw the digging of one canal, the repairing of a number of old dams, and the building of two new ones. In three of these colonies I tallied each day the additional number of trees cut for harvest. I saw many trees felled, and noted the manner in which they were moved by land and floated by water.

The greater number of the papers in this book were written especially for it. Parts of the others have been used in my books *Wild Life on the Rockies* and *The Spell of the Rockies*. "The Bea-

Preface

ver's Engineering" appeared in the *Saturday Evening Post*, and I am indebted to *McClure's* for permission to use "Beaver Pioneers."

Beaver works are of economical and educational value besides adding a charm to the wilds. The beaver is a persistent practicer of conservation and should not perish from the hills and mountains of our land. Altogether the beaver has so many interesting ways, is so useful, skillful, practical, and picturesque that his life and his deeds deserve a larger place in literature and in our hearts.

E. A. M.

Working like a Beaver

Working like a Beaver

ONE September day I saw a number of beaver at work upon a half-finished house. One part of the house had been carried up about two feet above the water, and against this were leaned numerous sticks, which stood upon the top of the foundation just above water-level. After these sticks were arranged, they were covered with turf and mud which the beaver scooped from the bottom of the pond. In bringing this earth covering up, the beaver invariably came out of the water at a given point, and over a short slide worn on the side of the house climbed up to the height where they were to deposit their load, which was carried in the fore paws. Then they edged round and put the mud-ball upon the house. From this point they descended directly to the water, but when they emerged with the next handful, they came out at the bottom of the slide, and again climbed up it.

The beaver often does a large amount of work

In Beaver World

in a short time. A small dam may be built up in a few nights, or a number of trees felled, or possibly a long burrow or tunnel clawed in the earth during a brief period. In most cases, however, beaver works of magnitude are monuments of old days, and have required a long time to construct, being probably the work of more than one generation. It is rare for a large dam or canal to be constructed in one season. A thousand feet of dam is the accumulated work of years. An aged beaver may have lived all his life in one locality, born in the house in which his parents were born, and he might rise upon the thousand-foot dam which held his pond and say, " My grandparents half a dozen centuries ago commenced this dam, and I do not know which one of my ancestors completed it."

Although the beaver is a tireless and an effective worker, he does not work unless there is need to do so. Usually his summer is a rambling vacation spent away from home. His longest period of labor is during September and October, when the harvest is gathered and general preparations made for the long winter. Baby beavers take part

Working like a Beaver

in the harvest-getting, though probably without accomplishing very much. During most winters he has weeks of routine in the house and ponds with nothing urgent to do except sleep and eat.

He works not only tooth and nail, but tooth and tail. The tail is one of the most conspicuous organs of the beaver. Volumes have been written concerning it. It is nearly flat, is black in color, and is a convenient and much-used appendage. It serves for a rudder, a stool, a prop, a scull, and a signal club. It may be used for a trowel, but I have never seen it so used. It serves one purpose that apparently has not been discussed in print; on a few occasions I have seen a beaver carry a small daub of mud or some sticks clasped between the tail and the belly. It gives this awkward animal increased awkwardness and even an uncouth appearance to see him humped up, with tail tucked between his legs, in order to clasp something between it and his belly.

He is accomplished in the use of arms and hands. With hands he is able to hold sticks and handle them with great dexterity. Like any clawing animal he uses his hands or fore paws, to dig

In Beaver World

holes or tunnels and to excavate burrows and water-basins. His hind feet are the chief propelling power in swimming, although the tail, which may be turned almost on edge and is capable of diagonal movement, is sometimes brought into play as a scull when the beaver is at his swiftest. In the water beaver move about freely and apparently with the greatest enjoyment. They are delightfully swift and agile swimmers, in decided contrast with their awkward slowness upon the ground. They can swim two hundred yards under water without once coming to the surface, and have the ability to remain under water from five to ten minutes. On one occasion a beaver remained under water longer than eleven minutes, and came to the top none the worse, apparently, for this long period of suspended breathing.

It is in standing erect that the beaver is at his best. In this attitude the awkwardness and the dull appearance of all-fours are absent, and he is a statue of alertness. With feet parallel and in line, tail at right angles to the body and resting horizontally on the ground, and hands held against the breast, he has the happy and childish eager-

A YOUNG BEAVER ON THE SIDE OF A BEAVER HOUSE

Working like a Beaver

ness of a standing chipmunk, and the alert and capable attitude of an erect and listening grizzly bear.

The beaver is larger than most people imagine. Mature male specimens are about thirty-eight inches in length and weigh about thirty-eight pounds, but occasionally one is found that weighs seventy or more pounds. Ten mature males which I measured in the Rocky Mountains showed an average length of forty inches, with an average weight of forty-seven pounds. The tails of these ten averaged ten inches in length, four and a half inches in width across the centre, and one inch in thickness. Behind the shoulders the average circumference was twenty-one inches, and around the abdomen twenty-eight. Ten mature females which I measured were only a trifle smaller.

There are twenty teeth; in each jaw there are eight molars and two incisors. The four front teeth of the beaver are large, orange-colored, strong, and have a self-sharpening edge of enamel. The ears are very short and rounded. The sense of smell appears to be the most highly developed

In Beaver World

of the beaver's senses. Next to this, that of hearing appears to be the most informational. The eyes are weak. The hind feet are large and webbed, and resemble those of a goose. The second claw of each hind foot is double, and is used in combing the fur and in dislodging the parasites from the skin. The fore paws of the beaver are handlike, and have long, strong claws. They are used very much after the fashion in which monkeys use their hands, and serve a number of purposes.

The color of the beaver is a reddish brown, sometimes shading into a very dark brown. Occasional specimens are white or black. The beaver is not a handsome animal, and when in action on the land he is awkward. The black skin which covers his tail appears to be covered with scales; the skin merely has this form and appearance, the scales do not exist. The tail somewhat resembles the end of an oar.

The all-important tools of this workman are his four orange-colored front teeth. These are edge-tools that are adaptable and self-sharpening. They are set in strong jaws and operated by

Working like a Beaver

powerful muscles. Thus equipped, he can easily cut wood. These teeth grow with surprising rapidity. If accident befalls them, so that the upper and the lower fail to bear and wear, they will grow by each other and in a short time become of an uncanny length. I have found several dead beaver who had apparently died of starvation; their teeth overlapped with jaws wide open and thus prevented their procuring food. For a time I possessed an overgrown tooth that was crescent-shaped and a trifle more than six inches long.

Pounds considered, the beaver is a powerful animal, and over a rough trail will drag objects of twice his own weight or roll a log-section of gigantic size. Up a strong current he will tow an eighty- or one-hundred-pound sapling without apparent effort. Three or four have rolled a one-hundred-and-twenty-pound boulder into place in the dam. Commonly he does things at opportune times and in the easiest way. His energy is not wasted in building a dam where one is not needed nor in constructive work in times of high water. He accepts deep water as a matter of fact and constructs dams to make shallow places deep.

In Beaver World

Beaver food is largely inner bark of deciduous or broad-leaved trees. Foremost among these trees which they use for food is the aspen, although the cottonwood and willow are eaten almost as freely. The bark of the birch, alder, maple, box-elder, and a number of other trees is also used. Except in times of dire emergency the beaver will not eat the bark of the pine, spruce, or fir tree. It is fortunate that the trees which the beaver fell and use for food or building purposes are water-loving trees, which not only sprout from both stump and root, but grow with exceeding rapidity. Among other lesser foods used are berries, mushrooms, sedge, grass, and the leaves and stalks of a number of plants. In winter dried grass and leaves are sometimes used, and in this season the rootstocks of the pond-lily and the roots of the willow, alder, birch, and other water-loving trees that may be got from the bottom of the pond. Beaver are vegetarians; they do not eat fish or flesh.

Apparently beaver prefer to cut trees that are less than six inches in diameter, and where slender poles abound it is rare for anything to be cut

Working like a Beaver

of more than four inches. But it is not uncommon to see trees felled that are from twelve to fifteen inches in diameter. In my possession are three beaver-cut stumps each of which has a greater diameter than eighteen inches, the largest being thirty-four inches. The largest beaver-cut stump that I have ever measured was on the Jefferson River in Montana, near the mouth of Pipestone Creek. This was three feet six inches in diameter.

The beaver sits upright with fore paws against the tree, or clasping it; half squatting on his hind legs, with tail either extending behind as a prop or folded beneath him as a seat, he tilts his head from side to side and makes deep bites into the tree about sixteen inches above the ground. In the overwhelming majority of beaver-cut trees that I have seen, most of the cutting was done from one side,—from one seat as it were. Though the notch taken out was rudely done, it was after the fashion of the axe-man. The beaver bites above and below, then, driving his teeth behind the piece thus cut off, will wedge, pry, or pull out the chip. Ofttimes in doing this he appears to

In Beaver World

use his jaw as a lever. With the aspen, or with other trees equally soft, about one hour is required to gnaw down a four-inch sapling. With one bite he will snip off a limb from half to three quarters of an inch in diameter.

After a tree is felled on land, the limbs are cut off and the trunk is gnawed into sections. The length of these sections appears to depend upon the size of the tree-trunk and also the distance to the water, the number of beaver to assist in its transportation, and the character of the trail. Commonly a six- or eight-inch tree is cut into lengths of about four to six feet. If the tree falls into the water of the pond or the canal, it is, if the limbs are not too long, transported butt foremost to the desired spot in its uncut, untrimmed entirety. Ofttimes with a large tree the trunk is left and only the limbs taken.

The green wood which the beaver uses for his winter's food-supply is stored on the bottom of the pond. How does he sink it to the bottom? There is an old and oft-repeated tale which says that the beaver sucks the air from the green wood so as to sink it promptly. Another tale has it that

Working like a Beaver

the beaver dives to the bottom carrying with him a green stick which he thrusts into the mud and it is thus anchored. Apparently the method is a simple one. The green wood stored is almost as heavy as water, and once in the pond it becomes water-logged and sinks in a short time; however, the first pieces stored are commonly large, heavy chunks, which are forced to the bottom by piling others on top of them. Frequently the first few pieces of the food-pile consist of entire trees, limbs and all. These usually are placed in a rude circle with butts inward and tops outward. This forms an entangling foundation which holds in place the smaller stuff piled thereon.

Most willows by beaver colonies are small and comparatively light. These do not sink readily, are not easily managed, and are rarely used in the bottom of the pile. Commonly, when these light cuttings are gathered into the food-pile, they are laid on top, where numerous up-thrusting limbs entangle and hold them. The foundation and larger portion of the food-pile are formed of heavy pieces of aspen, alder, or some other streamside tree, which cannot be moved out of place by an ordinary wind or water-current and which quickly sink to the bottom.

Among enemies of this fur-clad fellow are the wolverine, the otter, the lion, the lynx, the coyote, the wolf, and the bear. Hawks and owls occasionally capture a young beaver. Beaver spend much time dressing their fur and bathing, as they are harassed by lice and other parasites. At rare intervals they are afflicted with disease. They live from twelve to fifteen years and sometimes longer. Man is the worst enemy of the beaver.

A thousand trappers unite to tell the same pitiable tale of a trapped beaver's last moments. If the animal has not succeeded in drowning himself or tearing off a foot and escaping, the trapper smashes the beaver's head with his hatchet. The beaver, instead of trying to rend the man with sharp cutting teeth, raises himself and with upraised hand tries to ward off the death-blow. Instead of one blow, a young trapper frequently has to give two or three, but the beaver receives them without a struggle or a sound, and dies while vainly trying to shield his head with both hands.

Justly renowned for his industry, the beaver

Working like a Beaver

is a master of the fine art of rest. He has many a vacation and conserves his energies. He keeps his fur clean and his house in a sanitary condition. Ever in good condition, he is ready at all times for hard work and is capable of efficient work over long periods. He is ready for emergencies.

As animal life goes, that of the beaver stands among the best. His life is full of industry and is rich in repose. He is home-loving and avoids fighting. His lot is cast in poetic places.

The beaver has a rich birthright, though born in a windowless hut of mud. Close to the primeval place of his birth the wild folk of both woods and water meet and often mingle. Around are the ever-changing and never-ending scenes and silences of the water or the shore. Beaver grow up with the many-sided wild, playing amid the brilliant flowers and great boulders, in the piles of driftwood and among the fallen logs on the forest's mysterious edge. They learn to swim and slide, to dive quickly and deeply from sight, to sleep, and to rest moveless in the sunshine; ever listening to the strong, harmonious stir of wind

In Beaver World

and water, living with the stars in the sky and the stars in the pond; beginning serious life when brilliant clouds of color enrich autumn's hills; helping to harvest the trees that wear the robes of gold, while the birds go by for the southland in the reflective autumn days. If Mother Nature should ever call me to live upon another planet, I could wish that I might be born a beaver, to inhabit a house in the water.

Our Friend the Beaver

Our Friend the Beaver

ONE bright autumn afternoon I peered down into a little meadow by a beaver pond. This meadow was grass-covered and free from willows. In it seven or eight beaver were at work along a new canal. Each kept his place and appeared to have a section in which he did his digging. For more than half an hour I watched them clawing out the earth and grass-roots and lifting it out in double handfuls and piling it in an orderly line along the canal-bank. While I was watching a worker at one end of this line, two others clinched in a fight. The fighters made no sound except a subdued guttural mumbling as they rolled about in a struggle. The other workers, to my astonishment, paid not the slightest attention to this fight, but each attended to his own affairs. After two or three minutes the belligerents broke away; one squatted down breathing heavily, while the other, with bloody tail, dragged himself off and plunged into the pond.

In Beaver World

This was the first beaver fight that I had ever seen.

Beaver may well be called the silent workers. No matter how numerous, or crowded, or busy they are, their work goes on without a word and apparently without a sign. Although I have seen them at work scores of times, in the twilight and in the daylight, singly, in pairs, and by the dozens, doing the many kinds of work which beaver perform, yet this work has always gone quietly and without any visible evidences of management. Each one is capable of acting independently. Since the quality of his work improves as the beaver increases his experience, it appears natural and probable that each colony of beaver has a leader who plans and directs the work. I am familiar with a number of instances which strongly indicate leadership. In times of emergency, when an entire colony is forced to emigrate, a beaver — and usually an aged one — takes the lead, and wherever he goes the others willingly follow.

Whatever may have been the custom of beaver in the past, at present large numbers sometimes

Our Friend the Beaver

coöperate in accomplishing community work. It used to be believed, and possibly it was true, that only the members of a family, or the beaver of one house, united in doing the general work of the colony. It was a common belief that seven beaver inhabited a house; perhaps eight was the number of the Rocky Mountain region. At the present time the number in a house is from one to thirty.

Beaver have been driven from most of the streams and lake-shores, and now maintain themselves with difficulty in the places which they inhabit. In surviving they probably have had to sacrifice a few old customs and to adopt some new ones, and it is likely that these changes sometimes call for larger houses so as to care for the increased number of beaver which conditions now compel to live in one locality. A number of instances have come under my notice where beaver were driven from their colony either by fire or by the aggressiveness of trappers; these moved on to other scenes, where they cast their lot with the beaver of another colony, and apparently were received with every welcome. Immediately

In Beaver World

after the arrival of the immigrants, enlargements were at once commenced, apparently to accommodate the new-comers permanently.

One autumn, while following the Lewis and Clark trail with a pack horse in western Montana, I made camp one evening with a trapper who gave me a young beaver. He was about one month old, and ate twigs and bark as naturally as though he had long eaten them. I named him "Diver," and in a short time he was as chummy as a young puppy. Of an evening he played about the camp and often swam in the near-by water. At times he played at dam-building, and frequently displayed his accomplishment of felling wonderful trees that were about the size of a lead pencil. He never failed to come promptly when I whistled for him. At night he crouched near my camp, usually packing himself under the edge of the canvas on which I spread my bedding. Atop the pack on the horse's back he traveled, — a ride which he evidently enjoyed. He was never in a hurry to be taken off, and at moving time he was always waiting eagerly to be lifted on. As soon as he noticed me arranging

A YOUNG BEAVER SUNNING HIMSELF

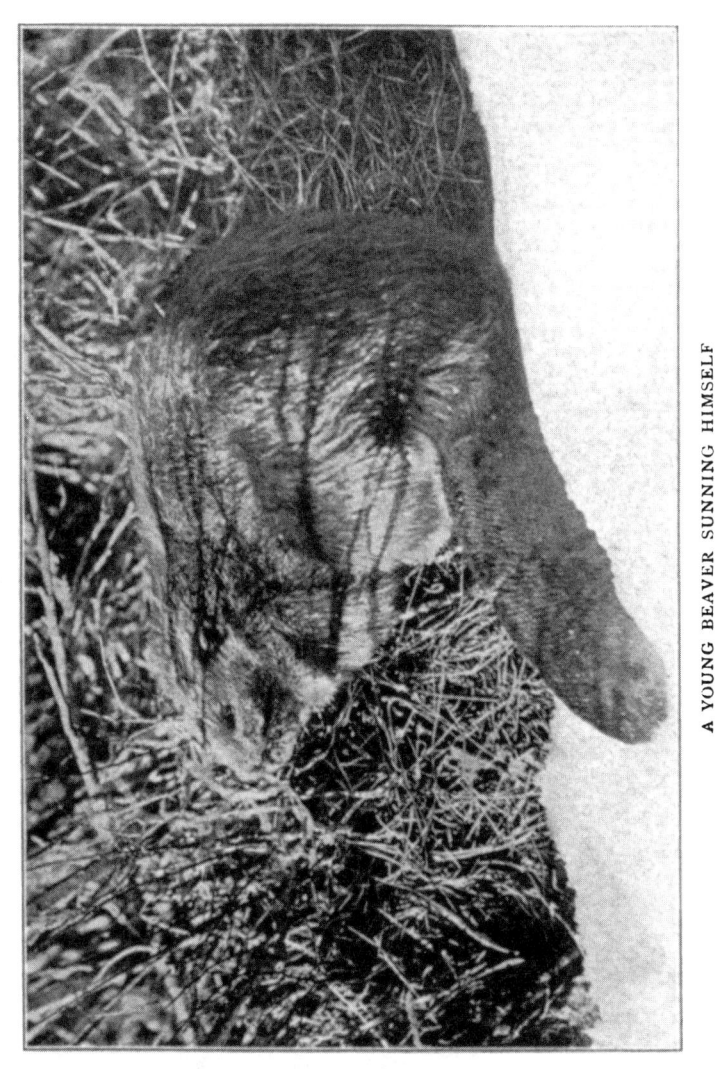

Our Friend the Beaver

the pack, he came close, and before I was quite ready for him, he rose up, extending his hands in rapid succession beggingly, and with a whining sort of muttering pleaded to be lifted at once to his seat on the pack.

He had a bad fright one evening. About one hour before sundown we had encamped as usual alongside a stream. He entered the water and after swimming about for a time, taking a dozen or so merry dives, he crossed to the opposite side. In plain view, only fifty feet away, I watched him as he busily dug out roots of the Oregon grape and then stopped leisurely to eat them. While he was thus engaged, a coyote made a dash for him from behind a boulder. Diver dodged, and the coyote missed. Giving a wail like a frightened child, my youngster rolled into the stream and dived. Presently he scrambled out of the water near me and made haste to crawl under my coat-tail behind the log on which I sat.

The nearest beaver pond was a quarter of a mile upstream, yet less than five minutes had elapsed from the time of Diver's cry when two beaver appeared, swimming low and cautiously

In Beaver World

in the stream before me. A minute later another came in sight from downstream. All circled about, swimming cautiously with heads held low in the water. One scented the place where the coyote had attacked Diver, and waddled out and made a sniffing examination. Another came ashore at the spot where Diver came out to me. Apparently his eyes told him I was a part of the log, but his nose proclaimed danger. After three or four hesitating and ineffectual attempts to retreat, he plucked up courage and rose to full height on hind legs and tail to stare eagerly at me. With head well up and fore paws drooping, he held the gaze for several seconds and then gave a low whistle.

At this, Diver came forth from behind my coat to see what was going on. The old one started forward to meet him, but on having a good look at me whirled and made a jumping dive into the water, whacking the surface with his tail as he disappeared. Instantly there followed two or more splashes and a number of tail-whacks upon the water, as though a beaver rescue party were beating a retreat.

Our Friend the Beaver

At the end of my outing Diver became the pet of two pioneer children on the bank of the Snake River. He followed the children about and romped with them. At three years of age he was shot by a visiting hunter.

My experience with Diver and other beaver pets leads me to believe that beaver are easily domesticated. One morning in northern Idaho, the family with whom I had spent the night took me out to see a beaver colony that was within a stone's throw of their fireplace. Three beaver came out of the water within ten feet of us to eat scraps of bread which the children threw on the grass for them.

One day I placed myself between three young beaver, who were eating on land, and the river out of which they came. They were on one of the rocky borders of the Colorado River in the depths of the upper Grand Cañon. They attempted to get by me, but their efforts were not of the "do or die" nature. Presently their mother came to the rescue and attempted to attract my attention by floating in the water near me in a terribly crippled condition. I had seen many

In Beaver World

birds and a few beaver try that clever ruse; so I allowed it to go on, hoping to see another act. Another followed.

In it an old male beaver appeared. He swam easily downstream until within a few yards of me and then dived, apparently frightened. But presently he reappeared near by and dived again. While I was watching him, the youngsters edged a few yards nearer the river. To stop them and prolong the exhibition, I advanced close to them as though to grab them. At this the mother beaver struggled out of the water and set up a tumbling and rolling so close to me that I thought to catch her for examination. She dodged right and left and reached the water. While this was going on, the youngsters escaped into the river. Mother beaver instantly recovered, and as she dived gave the water a scornful whack with her tail.

The beaver is not often heard. He works in silence. When he pauses from his work, he sits meditatively, like a philosopher. At times, however, when, in traveling, beaver are separated from one another, they give a strange shrill whistle or

Our Friend the Beaver

call. Occasionally this whistle appears to be a call of alarm, suspicion, or warning. Sometimes when alarmed, a young beaver gives a shrill and frightened cry not unlike that of a lost human child. On a few occasions I have heard, while listening near a beaver house in the early summer, something of a subdued concert going on inside, a purring, rhythmic melody. They have a kind of love ditty also. This is a rhythmic murmur and sigh, very appealing, and it seems strangely elemental as it floats across the beaver pond in the twilight.

It is probable that beaver mate for life. All that is known concerning their ways indicates that they are good parents. The young are usually born during the month of April. The number varies from one to eight; probably four is the number most common. A short time before the birth of the youngsters, the mother invites the father to leave, or compels him to do so,—or he may go voluntarily,—and she has possession of the house or burrow, probably alone, at the time the youngsters are born. Their eyes are open from the beginning, and in less than two weeks

In Beaver World

they appear in the water accompanied by the mother. Often I have investigated beaver colonies endeavoring to determine the number of youngsters at a birth. Many times there were four of these furry, serious little fellows near the house on a log that was thrust up through the water. At other times from one to eight youngsters sunned themselves on the top of the rude home.

One May, in examining beaver colonies, I saw three sets of youngsters in the Moraine Colony. They numbered three, and two, and five. One mother in another colony proudly exhibited eight, while still another, who had been harassed all winter by trappers and who lived in a burrow in the bank, could display but one.

It is not uncommon for young orphan beavers to be cared for and adopted by another mother beaver. I have notes of three mothers who, with children of their own, at once took charge of orphans left by the death of a neighbor. One June a mother beaver was killed near my camp. Her children escaped. The following evening a new mother, with four children of her own adopted them and moved from her own home, a quarter

Our Friend the Beaver

of a mile distant, to the home of her dead neighbor and there brought all the youngsters up.

Beaver have great fun while growing up. Posted on the edge of the house, they nose and push each other about, ofttimes tumbling one another into the water. In the water they send a thousand merry ripples to the shore, as they race, wrestle, and dive in the pond. They play on the house, in the pond, and in the sunshine and shadows of the trees along the shore.

Beaver are mature the third summer of their lives, and at this time they commonly leave the parental home, pair, and begin life for themselves. There are stories to the effect that the parents of the youthful home-builders accompany the children to new scenes, help them select a building-site, and assist in the construction of the new house and dam. After this the parents return home. This probably is occasionally true. Anyway I once saw this program fairly well carried out, and at another time in a limited manner.

The beaver is practical, peaceful, and industrious. He builds a permanent house and keeps it clean and in repair. Beside it he stores food-supply for the long winter. He takes thought for the morrow. These and other commendable characteristics give him a place of honor among the hordes of homeless, hand-to-mouth folk of the wild. During the winter he has but little to do except bathe and eat his two or three meals a day from the food he has stored in the autumn. Towards spring, when his wild neighbors are lean, hungry, and cold, he is fat and comfortable. In the spring he emerges from the house, but then his only work is occasionally to cut a twig for food. In the summer he plays tourist. He visits other colonies, and wanders up and down streams, going miles from home. In the late summer or early autumn he returns, makes repairs, and harvests food for winter.

The beaver is a valuable conservationist, but there are localities in which he cannot be tolerated. Although dead wood is rarely cut by the beaver, many a homesteader has been disturbed by his cutting off and carrying away green fence posts. Recently beaver have returned to a few localities and got themselves into bad repute by felling fruit trees. Occasionally, too, in the West,

Our Friend the Beaver

they have lost caste by persistently damming an irrigation-ditch and diverting the water, despite the fact that a court has given both the title and the right to this water to some one else a mile or so down the ditch.

In all logging operations, beaver never fail — where there is opportunity — to cut trees upstream and float them down with the current. Tree-cutting is an interesting phase of beaver life. A beaver will go waddling dully from the water to a tree he is about to cut down. All will look about for enemies; one may be wise enough — but the majority will not do so — to look upward to see if the tree about to be felled is entangled at the top. All appear to choose a comfortable place on which to squat or sit while cutting.

Commonly when the tree begins to creak and settle, the beaver who has done the cutting thuds the ground a few times with his tail, and then scampers away, usually going into the water. Sometimes the near-by workers give the thudding signal in advance of the one who is doing the cutting. Now and then no warning signal is given, and the logging beaver occasionally fells his tree upon other workers with a fatal result. As with axe-men, the beaver doing the cutting is on rare occasions caught and killed by the tree which he fells.

Rarely does the beaver give any thought to the direction in which the tree will fall. In a few instances, however, I have seen what appeared to be an effort on the part of the beaver to fell a tree in a given direction. From an uncomfortable place he cut the lowest notch on the side on which he probably wanted the tree to fall. On one of these occasions, the aspen tree selected stood in an almost complete circle of pines. The beaver took pains to cut the first and lowest notch in this tree directly opposite the opening in the pines. I have seen a number of instances of this kind. And he will sometimes leave the windward side of a grove on a windy day, and cut on the leeward, so that the felled trees are not entangled in falling.

Rarely does more than one beaver work at the same time at a tree. In some instances, however, if the tree be large, two or even more beaver will

IN THE HARVEST-FIELD
Aspens cut by beaver

Our Friend the Beaver

work at once. But after the tree has been felled, ofttimes three or four beaver will unite to roll a large section to the water. In doing this, some may stand with paws against it and push, and others may put their sides or hips against it. On land, as in the water, small limb-covered trees are dragged butt foremost so as to meet the least resistance. Sometimes the beaver drags walking backwards; at other times he is alongside the tree carrying and dragging it forward.

Early explorers say that beaver do most of their work at night. In this they are practically unanimous. However, in Long's Journal, written in 1820, beaver were reported at work in broad daylight. A few other early writers have also mentioned this daylight work. They probably work in darkness because that is the safest time for them to be out. During dozens of my visits to secluded localities, — localities which had not been visited by man, and certainly not by trappers, — I found beaver freely at work in broad daylight. I am inclined to think that day work was common during primeval times; and that, although the beaver now do and long have done most of

In Beaver World

their work at night, in localities where they are not in danger from man, they work freely during daytime.

Both the Indians and the trappers have a story that old beaver who will not work are driven from the colony and become morose outcasts, slowly living away the days by themselves in a burrow. I have no evidence to verify this statement, and am inclined to think that solitary beaver occasionally found in abandoned colony-sites and elsewhere are simply unfortunates, perhaps weighed down with age, unable to travel far, with teeth worn, the mate dead, without ambition to try, or without strength to emigrate. It is more likely that these aged ones voluntarily and sadly withdraw from their cheerful and industrious fellows, to spend their closing days alone. Although, too, there were among Indians and trappers stories of beaver slaves, I am without material for a story of this kind.

The beaver is peaceful. Although the males occasionally fight among themselves, the beaver avoids fighting, and plans his life so as to escape without it. Now and then in the water

Our Friend the Beaver

one closes with an otter in a desperate struggle, and when cornered on land one will sometimes turn upon a preying foe with such ferocity and skill that his assailant is glad to retreat. On two occasions I have known a beaver to kill a bobcat.

Beaver are not equally alert. In many cases this difference may be due to a difference in age or experience. Beaver have been caught with scars which show that they have been trapped before, a few even having lost two feet in escaping from traps. On the other hand, skillful trappers have found themselves after repeated trials, unable to catch a single beaver from a populous colony. Sometimes in colonies of this kind, the beaver even audaciously turned the traps upside down or contemptuously covered them with mud.

Nor is the work of all beaver alike. The ditches which one beaver digs, the house one builds, or the dam one makes, may be executed with much greater speed and with more skill than those of a neighboring beaver. Many houses are crude and unshapely masses, many dams haphazard in appearance, while a few canals are crooked and uneven. But the majority do good work, and are

In Beaver World

quick to take advantage of opportunity, quick to adjust themselves to new conditions, or to use the best means that is available. Beaver probably have made numbers of changes in their manners, habits, and customs, and those changes undoubtedly have enabled them to survive relentless pursuit, and to leave descendants upon the earth.

The industry of the beaver is proverbial, and it is to the credit of any person to have the distinction of working "like a beaver." Most people have the idea that the beaver is always at work; not that he necessarily accomplishes much at this work, but that he is always doing something. The fact remains that under normal conditions he works less than half the time, and it is not uncommon for him to spend a large share of each year in what might be called play. He is physically capable of intense and prolonged application, and, being an intelligent worker, even though he works less than half the time he accomplishes large results.

The Beaver Past and Present

The Beaver Past and Present

ALL Indian tribes in North America appear to have had one or more legends concerning the beaver. Most of these legends credit him with being a worthy and industrious fellow, and the Cherokees are said to trace their origin to a sacred and practical beaver. Many of the tribes had a legend which told that long, long ago the Great Waters surged around a shoreless world. These waters were peopled with beaver, beaver of a gigantic size. These, along with the Great Spirit, dived and brought up quantities of mud and shaped this into the hills and dales, the mountains where the cataracts plunged and sang, and all the caves and cañons. The scattered boulders and broken crags upon the earth were the missiles thrown by evil spirits, who in the beginning of things endeavored to hinder and prevent the constructive work of creation.

In Beaver World

The beaver has been found in fossil both in Europe and in America. Remnants of the dugout and the teeth of beaver, together with rude stone implements of primitive man, have been found in England. Near Albany, New York, gnawed beaver wood and the remains of a mastodon were dug up from about forty feet below the surface in sediment and river ooze. Fossil beaver were of enormous size.

Coming down to comparatively modern times, the animal as we now know him appears to have been distributed over almost all Asia, Europe, and North America. There was no marked difference in the individuals that inhabited these three continents. The beaver is probably extinct in Europe, but in July, 1900, I found a piece of wood floating in the Seine that had been recently gnawed by a beaver. At this time I was assured that not even a tame beaver could be found in Europe. It is still found in parts of Siberia and Central Asia. That form which inhabits South America is very unlike those in the Northern Hemisphere, and may be called a link between the muskrat and the beaver.

The Beaver Past and Present

Reference is made to the beaver in ancient Egyptian hieroglyphics, and Herodotus makes repeated mention of it. Pliny also gives a brief account of this animal. In Germany, in 1103, the right of hunting beaver was conferred along with other special hunting privileges; and a bull of Pope Lucius III, in 1182, gave to a monastery all the beaver found within the bounds of its property. A royal edict issued at Berlin in March, 1725, insisted upon the protection of beaver.

Before the white man came, beaver—*Castor canadensis*—were widely distributed over North America, perhaps more widely than any other animal. The beaver population was large, and probably was densest to the southwest of Hudson Bay and around the headwaters of the Missouri and Columbia rivers. Their scantiest population areas in the United States appear to have been southern Florida and the lower Mississippi Valley. This scantiness is attributed by early explorers to the aggressiveness of the alligators. All the southern half of Mexico appears to have been without a beaver population; but elsewhere over North America, wherever there were deciduous trees and water, and in a few treeless places where there were only water and grass, the beaver were found. Along the thousands of smaller streams throughout North America there was colony after colony, dam after dam, in close succession, as many as three hundred beaver ponds to the mile. Lewis and Clark mention the fact that near the Three Forks, Montana, the streams stretched away in a succession of beaver ponds as far as the eye could reach. The statements made by the early explorers, settlers, and trappers, together with my own observations,— which commenced in 1885, and which have extended pretty well over the country from northern Mexico into Alaska,— lead to the conclusion that the beaver population of North America at the beginning of the seventeenth century was upwards of one hundred million. The area occupied was approximately six million square miles, and probably two hundred beaver population per square mile would be a conservative number for the general average.

In the United States there are a number of counties and more than one hundred streams and lakes named for the beaver; upwards of fifty post-

The Beaver Past and Present

offices are plain Beaver, Beaver Pond, Beaver Meadow, or some other combination that proclaims the former prevalence of this widely distributed builder. The beaver is the national emblematic animal of Canada, and there, too, numerous post-offices, lakes, and streams are named for the beaver.

Beaver skins lured the hunter and trapper over all American wilds. These skins were one of the earliest mediums of exchange among the settlers of North America. For two hundred years they were one of the most important exports, and for a longer time they were also the chief commodity of trade on the frontier. A beaver skin was not only the standard by which other skins were measured in value, but also the standard of value by which guns, sugar, cattle, hatchets, and clothing were measured. Though freely used by the early settlers for clothing, they were especially valuable as raw material for the manufacture of hats, and for this purpose were largely exported.

From this animal were prepared many remedies which in former times were believed to have high medicinal value. Castoreum was the most

In Beaver World

popular of these, and from it was compounded the great cure-all. The skin of the beaver was thought to be an excellent preventive of colic and consumption; the fat of the beaver efficient in apoplexy and epilepsy, to stop spasms, and for various afflictions of the nerves. Powdered beaver teeth were often given in soup for the prevention of many diseases. The castoreum of the beaver was considered a most efficient remedy for earache, deafness, headache, and gout, for the restoring of the memory and the cure of insanity. Next in importance to its skin, the beaver was valued for the castoreum it yielded.

The old hunters, trappers, and first settlers forecast with confidence the weather from the actions of the beaver. This animal was credited with being weather-wise to a high degree. From his actions the nature of the oncoming winter was predicted, and plans to meet it were made accordingly. Faith in the beaver's actions and activities as a basis for weather-forecasting was almost absolute. If the beaver began work early, the winter was to begin early. If the beaver laid up a large harvest, covered the house deeply with mud, and

The Beaver Past and Present

raised the water-level of the pond, the winter was, of course, to be a long and severe one.

Extensive autumn rambles in the mountains with especial attention to beaver customs compels me to conclude that as a basis for weather prediction beaverdom is not reliable. In the course of one autumn month in the mountains of Colorado more than one hundred colonies were observed. In many colonies work for the winter commenced early. In others, only a few miles distant, preparations for the winter did not begin until late. In some, extensive preparations were made for the winter. In a few the harvest laid up was exceedingly small. Thus in one month of the same year I saw some beaver colonies preparing for a long winter and others for a short one, many preparing for a hard winter and others almost unprepared for winter. From these varied and conflicting prognostications, how was one accurately to forecast the coming winter? The old prophets in one colony frequently disagreed with aged prophets who were similarly situated, but in a neighboring colony. At one place thirty or more beaver gathered an enormous quantity of food, sufficient, in fact, to have supplied twice that number for the longest and most severe winter. The winter which followed was as mild a one as had passed over the Rocky Mountains in fifty years. Not one tenth of the big food-pile was eaten.

I have not detected anything that indicates that the beaver ever plan for an especially hard winter. Goodly preparations are annually made for winter. Apparently the extent of the preparation in any colony is dependent almost entirely upon the number of beaver that are to winter in that colony. Winter preparations consist of gathering the food-harvest, repairing and sometimes raising the dam, and commonly covering the house with a layer of mud. Beaver display forethought, intelligence, and even wisdom, but being weather-wise is not one of their successful specialties. Local beaver now and then show unusual activity, and unusually large supplies are gathered and stored for the winter. This kind of work appears to be local, not general. The cases in which unusually large preparations were made for the winter could have been traced to an increased

The Beaver Past and Present

population of the colony that showed these activities. On the other hand, colonies with less preparations one year than on the preceding one probably had suffered a decrease of population. Increase of population in a beaver colony may be accounted for through the growing up of youngsters, or by the arrival of immigrants, or both; where the temporary inactivity of trappers in one locality might allow the beaver colony in that region to increase in numbers; or where the beaver population of that colony might be increased by the arrival of beaver driven from their homes by aggressive hunters and trappers in adjoining localities. At any rate, in the beaver world, some colonies each year commence work earlier than do others, and some colonies make extensive preparations for the winter, while others make but little preparation. This preparation appears to be determined chiefly by the number of colonists and the needs of the colony.

The beaver hastened, if it did not bring, the settlement of the country. Hunters and trappers blazed the trails, described the natural resources, and lured the permanent settlers to possess the

In Beaver World

land and build homes among the ruins left by the beaver. Early in the fur industry companies were formed, the Hudson's Bay Company becoming the most influential and best known. Its charter was granted by Charles II of England on the 2d day of May, 1669. This company finally developed into one of the greatest commercial enterprises that America has ever known. The skin of the beaver furnished more than half its revenue. There are many features in the history of this company that have never been surpassed in any land. For more than two hundred years it held absolute sway over a country larger than Europe, and for the first one hundred and fifty years of its existence it was the government of the territory where it ruled, and thus determined the social and other standards of life within that territory. One of the early officials of this company declared that they were on the ground ahead of the missionaries, and said that the initials "H. B. C." on the banner of the company might well be interpreted as "Here before Christ."

Kingsford's History of Canada says that in the eighteenth century, Canada exported a moderate

The Beaver Past and Present

quantity of timber, wheat, the herb called ginseng, and a few other commodities, but from first to last she lived chiefly on beaver skins. Horace T. Martin, formerly Secretary of Agriculture for Canada, calls the beaver's part in Canadian development "a subject which has from the inception of civilization been associated with the industrial and commercial development, and indirectly with the social life, the romance, and to a considerable extent with the wars of Canada."

The American Fur Company and the Northwestern Fur Company were two large fur-gathering enterprises whose trappers ranged afar and who left their mark in the history and the development of the Northwest. The colossal Astor fortune really had its beginning in the wealth which John Jacob Astor amassed chiefly through the gathering and the sale of beaver skins. Beaver skins are now economically unimportant in commerce, but their value has already led to the establishment of a few beaver farms.

To-day beaver are apparently extinct over the greater portion of the area which they formerly occupied, and are scarce over the remaining inhabited area. Scattered colonies are found in the Rocky Mountains and in the mountains of the Pacific Coast, and there are localities in Canada where they are still fairly abundant. In many places in the Grand Cañon of the Colorado they are common. A few are found in Michigan and Maine. Some years ago a few brooks in the Adirondacks were successfully colonized with these useful animals. They have reappeared in Pennsylvania, and there probably are straggling beaver all over the United States which, if protected, would increase.

There is a growing sentiment in favor of allowing the beaver to multiply. In 1877 Missouri passed a law protecting these animals; so did Maine in 1885 and Colorado in 1899. Other States to the total number of twenty-four have also legislated for their protection. The Canadian government has also passed protective laws. A noticeable increase has already occurred in a few localities. Beaver multiply rapidly under protection, as is shown in the National Parks of both Canada and the United States.

As Others See Him

As Others See Him

FOR three hundred years the beaver has been a popular subject for discussion. Fabulous accounts have been given concerning his works, and that which he has done has been exaggerated beyond recognition. Many of the descriptions of him are grotesque, and many accounts of his works are uncanny. His tail has been made to do the work of a pile-driver, and some of the old accounts credit him with driving stakes into the ground that were as large as a man's thigh and five or six feet long. Stories have been told that his tail was used as a trowel in plastering the house and the dam. A few writers have stated that he lived in a three-story lodge. More than a century ago Audubon called attention to the enormous mass of fabrications that had been written concerning this animal, and in 1771 Samuel Hearne of the Hudson's Bay Company denounced a beaver nature-faker in the following terms: "The compiler of the

In Beaver World

Wonders of Nature and Art seems to have not only collected all the fictions into which other writers on the subject have run, but has so greatly improved on them that little remains to be added to his account of the beaver beside a vocabulary of their language, a code of their laws, and a sketch of their religion, to make it the most complete natural history of that animal."

One might read almost the entire mass of printed matter concerning the beaver without obtaining correct information about his manners and customs or an accurate description of his works and without getting at the real character of this animal. The actual life and character of the beaver, however, the work which he does, the unusual things which he has accomplished, are really more interesting and place the beaver on a higher plane than do all the fictitious tales and exaggerated accounts written concerning him.

Mr. Lewis H. Morgan in his "American Beaver and his Works" says: "No other animal has attracted a larger share of attention or acquired by his intelligence a more respectable position in the public estimation. Around him are the dam,

As Others See Him

the lodge, the burrow, the tree-cutting, and the artificial canal, each testifying to his handiwork, and affording us an opportunity to see the application as well as the results of his mental and physical powers. There is no animal below man in the entire range of Mammalia which offers to our investigation such a series of works, or presents such remarkable material for study and illustration of animal psychology."

Mr. Morgan was for years a capable and painstaking student of the beaver. That which he has written is so important a contribution concerning the beaver that no one interested in this animal can afford to be unacquainted with it. In the preface of his book he says: " I took up the subject as I did fishing, for summer recreation. In the year 1861, I had occasion to visit the Red River Settlement in the Hudson's Bay Territory, and in 1862, to ascend the Missouri River to the Rocky Mountains, which enabled me to compare the works of the beaver in these localities with those on Lake Superior. At the outset I had no expectation of following up the subject year after year, but was led on, by the interest

In Beaver World

which it awakened, until the materials collected seemed to be worth arranging for publication."

The greatest admirers of the beaver are those who know him best. He bears acquaintance. This cannot be had by merely looking at the animal, nor by sympathetically studying his monumental works. These works will of course impress one, but they give one at best only a traveler's impression. Long and repeated visits to the colony in its busy season appear to be the best way to get at the character of the beaver. The cubical contents of a dam may not even suggest the obstacles overcome in its construction, the labor of getting the material, the dangers avoided, the numerous unexpected difficulties overcome. Five cords of green poles and limbs in a neat pile in the pond by the beaver house may tell that the harvest has been gathered, but it does not tell that a part of this harvest may have been gathered a mile away and skillfully transported to the house with difficulty and amid dangers. A part of the food-pile may have been dragged laboriously uphill and along trails which required months of labor to open; or numerous

As Others See Him

pieces in this pile may have been floated through a canal of such magnitude that a generation was required to construct it. Altogether, harvest-gathering is interesting and heroic work on the part of the beaver. In doing it he takes large risks, for the harvest is usually gathered far from the house and on the dangerous beaver frontier.

For more than a quarter of a century I have been a friendly visitor to his colonies, in which I have lingered long and lovingly. That he makes mistakes is certain, but that he is an intelligent, reasoning animal I have long firmly believed. As I said in "Wild Life on the Rockies," — "I have so often seen him change his plans so wisely and meet emergencies so promptly and well that I can think of him only as a reasoner."

As evidence that he sometimes reasons, it may be cited that he occasionally endeavors to fell trees in a given direction; that he often avoids cutting those entangled at the top; that sometimes he will, on a windy day, fell trees on the leeward side of a grove; that he commonly avoids felling trees in the heart of a grove, but cuts on the outskirts of it. He occasionally dams a stream, digs a canal, leads water to a dry place, and there forms and fills a reservoir and establishes a home. Often his house is built by a spring and thus the danger from thick ice avoided. These are some of the reasons for my believing him to be intelligent.

Morgan speaks of the beaver as "endowed with a mental principle which performs for him the same office that the human mind does for man," and says, "The works of the beaver afford many interesting illustrations of his intelligence and reasoning capacity," also, "In the capacity thereby displayed of adapting their works to the ever-varying circumstances in which they find themselves placed instead of following blindly an invariable type, some evidence of possession on their part of *free intelligence* is undoubtedly furnished."

Mr. George J. Romanes has the following opinion of the beaver: "Most remarkable among rodents for instinct and intelligence, unquestionably stands the beaver. Indeed there is no animal — not even excepting the ants and bees — where instinct has risen to a higher level of far-

As Others See Him

reaching adaptation to certain constant conditions of environment, or where faculties, undoubtedly instinctive, are more puzzlingly wrought up with faculties no less undoubtedly intelligent. . . . It is truly an astonishing fact that animals should engage in such vast architectural labors with what appears to be the deliberate purpose of securing, by such artificial means, the special benefits that arise from their high engineering skill. So astonishing, indeed, does this fact appear, that as sober minded interpreters of fact we would fain look for some explanation which would not necessitate the inference that these actions are due to any intelligent appreciation, either of the benefits that arise from labor, or of the hydrostatic principles to which this labor so clearly refers."

Mr. Alexander Majors, originator of the Pony Express, who lived a long, alert life in the wilds, pays the beaver the following peculiar tribute in his "Seventy Years on the Frontier": "The beaver, considered as an engineer, is a remarkable animal. He can run a tunnel as direct as the best engineer could do with his instruments to guide him. I have seen where they have built a dam across a stream, and not having sufficient head water to keep their pond full, they would cross to a stream higher up the side of the mountain, and cut a ditch from the upper stream and connect it with the pond of the lower, and do it as neatly as an engineer with his tools could possibly do it. I have often said that the beaver in the Rocky Mountains had more engineering skill than the entire corps of engineers who were connected with General Grant's army when he besieged Vicksburg on the banks of the Mississippi. The beaver would never have attempted to turn the Mississippi into a canal to change its channel without first making a dam across the channel below the point of starting the canal. The beaver, as I have said, rivals and sometimes even excels the ingenuity of man."

Longfellow translates the spirit of the beaver world into words, and enables one in imagination to restore the primeval scenes wherein the beaver lived: —

> "Should you ask me, whence these stories?
> Whence these legends and traditions,

As Others See Him

With the odors of the forest,
With the dew and damp of meadows,

. . . .

Should you ask where Nawadaha
Found these songs so wild and wayward,
Found these legends and traditions,
I should answer, I should tell you,
' In the bird's-nests of the forest,
In the lodges of the beaver.' "

And the cunning Pau-Puk-Keewis, fleeing from the wrath of Hiawatha, ran, —

" Till he came unto a streamlet
In the middle of the forest,
To a streamlet still and tranquil,
That had overflowed its margin,
To a dam made by the beavers,
To a pond of quiet water,
Where knee-deep the trees were standing,
Where the water-lilies floated,
Where the rushes waved and whispered.
 On the dam stood Pau-Puk-Keewis,
On the dam of trunks and branches,
Through whose chinks the water spouted,
O'er whose summit flowed the streamlet.
From the bottom rose the beaver,
Looked with two great eyes of wonder,
Eyes that seemed to ask a question,
At the stranger Pau-Puk-Keewis."

The Beaver Dam

The Beaver Dam

MILLIONS of beaver ponds graced America's wild gardens at the time the first settlers came. These ragged and poetic ponds varied in length from a few feet to one mile, and in area they were from one hundred acres down to a miniature pond that half a dozen merry children might encircle. These ponds were formed by dams built by beaver, and the dams varied greatly in size and were made of poles variously combined with sticks, stones, trash, rushes, and earth.

In the Bad Lands of Dakota I saw two dams that were made of chunks of coal. This material had caved from a near-by bluff. I have noticed a few that were constructed of cobble-stones. The water-front of these dams was filled and covered with clay, and they were the work of "grass beavers,"—beaver that subsist chiefly on grass, and that live in localities almost destitute of trees.

It is doubtful if a dam is ever made by felling

In Beaver World

logs or large trees across the stream. I have, however, seen a few real log dams, but in these the logs were placed parallel to the flow of water. One of these was in the Sawtooth Mountains of Idaho. Here a snow-slide swept several hundred trees down the mountain. This wreckage was piled on the bank of a stream. Beaver in a colony a short distance away accepted this gift of the gods, and of these unwieldy logs built a dam about two hundred feet downstream from where the avalanche had piled the logs. This dam was a massive affair, about forty feet long and eight feet high. It really appeared more like a log jam than a dam, but it served the purpose intended and raised the level of the river so that the water overflowed to one side and spread in a broad sheet against a cliff and through a grove of aspens, which the beaver proceeded to harvest.

The majority of dams are made of slender green poles which are placed lengthwise with the flow for the bottom, and set braced with the end upstream a foot or so higher than the downstream end. With these there are occasionally used small limby trees. The large end of the tree

A NEW DAM

The Beaver Dam

is placed upstream, and the small bushy end downstream. If in a current these sometimes are weighed down with mud or stones. Short, stout sticks and long, slender poles are deftly mingled in the dam as it rises. The poles overlie, and many completed dams appear as though made of gigantic inclined half-closed shears and compasses of poles. Thus a dam is doubly braced. The weight against it is resisted both by the end-on poles that are parallel to the flow and by those set at an angle to it.

The shape and the material of a dam are dependent on a number of things: the nature of the place where built, the kind of materials available for its building, the purpose it is intended to serve, and the relation it may have to dams already constructed. Sometimes a small dam will be made — that may ultimately become a big one — by simply digging a ditch across the stream or basin and piling the excavated material into a dam.

Beaver, like men, are unequal in their skill, both in planning and in doing work, and the work of most beaver falls short of perfection. Errors are not uncommon. More than one colony has commenced a dam apparently without knowing that there was not sufficient available material to complete it. Others have built in the wrong places, and have thus failed to flood the area which they desired to reach or cover with water. Occasionally the difficulties of construction have been too great for the beaver who attempted it, and the dam has been abandoned in an incomplete state. Now and then a weak dam breaks, or a strong one is swept out by a flood.

But why do beaver need or want the pond which the dam forms? They need it for the purpose of maintaining water of sufficient depth and area to enable them to move about in safety, and to transport their food-supplies with the greatest ease. Above all, the pond is a place of refuge into which the beaver can constantly plunge and have security from his numerous and ever watchful enemies. The house-entrance must be kept water-covered. In the water the beaver is in his element. On the land he is a child lost in the wilds. He has extremely short legs and a heavy body. His make-up fits him for movement in the water. He is a

The Beaver Dam

graceful swimmer, and in the water can move easily and evade enemies; while on land he is an awkward lubber, moves slowly, and is easily overtaken. Water of sufficient depth and area, then, is essential to the life and happiness of the beaver. To have this at all times it is necessary, in localities where the supply is at times insufficient, to maintain it by means of dams and ponds.

Deep ponds are needed around the house; shallow ponds with shores in near-by groves facilitate far-away logging. Dams are placed across streams whose waters are to be led away through new channels and made to serve elsewhere in canals or ponds. Dams are made across inclined canals to catch and hold water in them. Streams are beaver's avenues of travel. Along shallow streams in a beaver country it is not uncommon to see an occasional short dam which forms a deep hole, which apparently is maintained as a harbor or place of safety into which traveling beaver may dive and be made safe from pursuit.

Most beaver dams are built on the installment plan. They are the result of growth. The new

In Beaver World

dam is short and comparatively low. It is enlarged as conditions may require. As the trees in the edge of the pond are harvested, the dam is built higher and longer, so as to flood a larger area; or as sediment fills the pond, the dam is from time to time raised and lengthened in order to maintain the desired depth of water. Thus it may grow through the years until the possibilities of the locality are exhausted. The dam may then be abandoned. It may be used for a few years or it may be used for a century. A gigantic beaver dam may thus represent the work of several generations of beaver. It often occurs that one or more generations may use a dam and yearly add something to its size. By and by these beaver may die or emigrate. The old dam remains, falling to ruin in places. Years go by and other beaver come upon the scene. The old dam is then used for the foundation for a new one. The appearance of some old dams indicates that they have been repeatedly used and abandoned.

New dams, being made largely of coarse materials, appear very unlike old ones. Decay, settling, repairs, and other changes come rapidly. The

The Beaver Dam

dam is built of poles to-day; it speedily becomes earthy and is planted by nature to grass, willows, and flowers. On old, large dams it is not uncommon to see old forest-trees. The roots of these entangle the constructive materials, penetrate deeply, and help to anchor securely the entire dam.

In only a few cases are the water-fronts of dams at once plastered or filled in with mud. This is done only where there is a scarcity of water. It is the aim of the beaver to raise the water in the pond to a certain height and there maintain it, the chief purpose of the dam being to regulate the height or the depth of the water. The water, in streaming through new dams, deposits therein quantities of sticks, trash, and sediment, so that in a year or two these choke the holes, almost stop the leakage of the water, and help to solidify the dam. The discharge from dams is regulated by the beaver. In some instances water leaks through a dam in numerous places from bottom to top; in others it seeps through only close to the top; and in still others the dam is so solid that the water pours over the top in a thin sheet. In some cases, however, instead of the water pouring over the entire length of the dam the beaver force it to pour over in a given stretch at one end or the other, or sometimes through a hole or tunnel. The concentration of the overflow at some one point in the dam is commonly done either for the purpose of using it in transportation or to force the water to outpour on a spot where it will least erode the foundation of the dam. Occasionally beaver compel the water to flow round the end of a dam, which they raise sufficiently high for that purpose. Sometimes they dig a waste-way for the water.

European beaver appear to have barely developed to the dam-building stage. Rarely did they build even a small, unimportant dam. Nor did all the American beaver build dams. At the time the beaver population was most numerous and widely distributed, probably not more than half of them used the dam. However, those not using the dam were living in places where the dam and consequent pond were not needed. Dam-building enormously increased the habitable beaver area. There were, and are, thousands of brooks which

The Beaver Dam

each year cease to flow for a period, yet on these brooks are all other beaver requirements except a permanent, sufficient water-supply. By dam-building water is stored for to-morrow, or stream-courses changed, and with the assistance of canals water is diverted to a dry ravine where a colony is established.

The dam is the largest and in many respects the most influential beaver work. Across a stream it is an inviting thoroughfare for the folk of the wild. As soon as a dam is completed, it becomes a wilderness highway. It is used day and night. Across it go bears and lions, rabbits and wolves, mice and porcupines; chipmunks use it for a bridge, birds alight upon it, trout attempt to leap it, and in the evening the graceful deer cast their reflections with the willows in its quiet pond. Across it dash pursuer and pursued. Upon it take place battles and courtships. Often it is torn by hoof and claw. Death struggles stain it with blood. Many a drama, romantic and picturesque, fierce and wild, is staged upon the beaver dam.

The beaver dam gives new character to the

landscape. It frequently alters the course of a stream and changes the topography. It introduces water into the scene. It nourishes new plant-life. It brings new birds. It provides a harbor and a home for fish throughout the changing seasons. It seizes sediment and soil from the rushing waters, and it sends water through subterranean ways to form and feed springs which give bloom to terraces below. It is a distributor of the waters; and on days when dark clouds are shaken with heavy thunder, the beaver dam silently breasts, breaks, and delays the down-rushing flood waters, saves and stores them; then, through all the rainless days that follow, it slowly releases them.

Most old colonies have many dams and ponds. A dam is sometimes built for the purpose of forcing water back and to one side into a grove that is to be harvested for food. In many cases water flows round the end of a dam, and in making its way back to the main channel is intercepted by another dam, then another; and thus the water from one small brook maintains a cluster or chain of pondlets.

The Beaver Dam

The majority of beaver dams are as crooked as a river's course. Now and then one is straight. A few are built from shore to a boulder, from the boulder to a willow-clump, and finally, perhaps, from willow-clump to some outstretching peninsula on the further shore. It is not uncommon for a short dam to be built and afterwards lengthened with additions on each end which may curve either down or up stream. Sometimes a dam is built outward from opposite shores simultaneously by separate but coöperating crews of beaver. In swift water these ends are forced downstream in building, so that when they are finally joined midstream the dam curves noticeably downstream.

On one occasion I watched beaver commence and complete a dam in moderately swift water that when finished bowed strongly *upstream.* This, however, was not the intention of the builders. The material for this dam consisted of willow and alder poles that were cut some distance upstream. These were floated down as used. This dam was begun against a huge boulder near midstream, and built outward simultaneously to-

In Beaver World

ward both shores. Despite the repeated efforts of the builders to extend it in a straight line to the shore, the flow of the water pushed these outbuilding ends downward, and when they finally reached the shore this fifty-odd feet of dam with the boulder for a keystone had an arch that was about fifteen feet in advance of the bases.

Not far from where I lived in the mountains when a boy, the beaver built a dam. This had a slight arch upstream. A few years later the dam was doubled in length by building an extension on the end which bowed downstream. It thus stood a reverse curve. Later the dam was still further lengthened by a comparatively straight stretch on one end, and by a short, down-bowing stretch on the other. Recent additions to this dam consist of wings at the end which sweep upstream. The dam as it now stands reaches about three fourths of the way around the pond which it forms.

It is not uncommon for a dam to be planned and built with an arch against the current or against the water which it afterwards impounds. The most interesting dam of this kind that I ever

The Beaver Dam

saw was one across the narrow neck of a rudely bell-shaped basin that was about two hundred feet in length. The material for this dam came from a grove of aspens that extended into one side of the basin. The floor of this basin was partly covered with a few inches of water. In starting the dam the beaver evidently knew where they wanted to build it. This was not by the aspen grove where the materials were convenient, where the dam would need to be about one hundred and twenty feet long, but was about fifty feet farther on, where a dam of only forty feet was required. This dam when completed bowed seven feet against the enclosed water. The beaver commenced building at the end nearest the grove of aspens, pulling and dragging the poles the fifty feet to it. They laid these aspen poles, which were two to five inches in diameter and from four to twelve feet in length, at right angles to the length of the dam, and usually placed the large end upstream or against the current. But the water was shallow, and the transportation of these poles to the dam was difficult. Accordingly a ditch or canal was dug from the

In Beaver World

grove to the place by the dam where the work was going on. This ditch was about twenty-five inches wide and fifteen deep. The waters filled it and thereby afforded an easy means of floating or transporting the poles from the grove to the place where they were being used. This ditch was carried forward along the upper line of the dam, and several feet in advance of the spot where the outbuilding work was advancing. Upon the earth thrown up from this were laid the upper or high ends of the poles. When the dam was finally completed, it was approximately eight feet wide on the base and stood four feet high. As soon as it was completed, the beaver stuffed the water-front with mud and grass roots, which were obtained by digging from the construction ditch immediately in front of the dam. In other words, they enlarged their pole-floating ditch above the dam into a deeper and wider channel, and used this excavated material for strengthening and waterproofing the dam.

The longest beaver dam that I have ever seen or measured was on the Jefferson River near

The Beaver Dam

Three Forks, Montana. This was 2140 feet long. Most of it was old. More than half of it was less than six feet in height; two short sections of it, however, were twenty-three feet wide at the base, five on top, and fourteen feet high.

Harvest Time with Beavers

Harvest Time with Beavers

ONE autumn I watched a beaver colony and observed the customs of its primitive inhabitants as they gathered their harvest for winter. It was the Spruce Tree Colony, the most attractive of the sixteen beaver municipalities on the big moraine on the slope of Long's Peak.

The first evening I concealed myself close to the beaver house by the edge of the pond. Just at sunset a large, aged beaver of striking, patriarchal appearance rose in the water by the house, and swam slowly, silently round the pond. He kept close to the shore and appeared to be scouting to see if an enemy lurked near. On completing the circuit of the pond, he climbed upon the end of a log that was thrust a few feet out into the water. Presently several other beaver appeared in the water close to the house. A few of these at once left the pond and nosed quietly about on the shore. The others swam about for some minutes and then joined their comrades on land, where all rested for a time.

Meanwhile the aged beaver had lifted a small aspen limb out of the water and was squatted on the log, leisurely eating bark. Before many minutes elapsed the other beavers became restless and finally started up the slope in a runway. They traveled slowly in single file and one by one vanished amid the tall sedge. The old beaver slipped noiselessly into the water, and a series of low waves pointed toward the house. It was dark as I stole away in silence for the night, and Mars was gently throbbing in the black water.

This was an old beaver settlement, and the numerous harvests gathered by its inhabitants had long since exhausted the near-by growths of aspen, the bark of which is the favorite food of North American beaver, though the bark of the willow, cottonwood, alder, and birch is also eaten. An examination of the aspen supply, together with the lines of transportation, — the runways, canals, and ponds, — indicated that this year's harvest would have to be brought a long distance. The place it would come from was an aspen

Harvest Time with Beavers

grove far up the slope, about a quarter of a mile distant from the main house, and perhaps a hundred and twenty feet above it. In this grove I cut three notches in the trunks of several trees to enable me to identify them whether in the garnered pile by a house or along the line of transportation to it.

The grounds of this colony occupied several acres on a terraced, moderately steep slope of a mountain moraine. Along one side rushed a swift stream on which the colonists maintained three but little used ponds. On the opposite side were the slope and summit of the moraine. There was a large pond at the bottom, and one or two small ponds, or water-filled basins, dotted each of the five terraces which rose above. The entire grounds were perforated with subterranean passageways or tunnels.

Beaver commonly fill their ponds by damming a brook or a river. But this colony obtained most of its water-supply from springs which poured forth abundantly on the uppermost terrace, where the water was led into one pond and a number of basins. Overflowing from these, it either made a

In Beaver World

merry little cascade or went to lubricate a slide on the short slopes which led to the ponds on the terrace below. The waters from all terraces were gathered into a large pond at the bottom. This pond measured six hundred feet in circumference. The crooked and almost encircling grass-grown dam was six feet high and four hundred feet long. In its upper edge stood the main house, which was eight feet high and forty feet in circumference. There was also another house on one of the terraces.

After notching the aspens I spent some time exploring the colony grounds and did not return to the marked trees until forty-eight hours had elapsed. Harvest had begun, and one of the largest notched trees had been felled and removed. Its gnawed stump was six inches in diameter and stood fifteen inches high. The limbs had been trimmed off, and a number of these lay scattered about the stump. The trunk, which must have been about eighteen feet long, had disappeared, cut into lengths of from three to six feet, probably, and started toward the harvest pile. Wondering for which house these logs were

Harvest Time with Beavers

intended, I followed, hoping to trace and trail them to the house, or find them *en route*. From the spot where they were cut, they had evidently been rolled down a steep, grassy seventy-foot slope, at the bottom of this dragged an equal distance over a level stretch among some lodgepole pines, and then pushed or dragged along a narrow runway that had been cut through a rank growth of willows. Once through the willows, they were pushed into the uppermost pond. They were taken across this, forced over the dam on the opposite side, and shot down a slide into the pond which contained the smaller house. Only forty-eight hours before, the little logs which I was following were in a tree, and now I expected to find them by this house. It was good work to have got them here so quickly, I thought. But no logs could be found by the house or in the pond! The folks at this place had not yet laid up anything for winter. The logs must have gone farther.

On the opposite side of this pond I found where the logs had been dragged across the broad dam and then heaved into a long, wet slide which landed them in a small, shallow harbor in the grass. From this point a canal about eighty feet long ran around the brow of the terrace and ended at the top of a long slide which reached to the big pond. This canal was new and probably had been dug especially for this harvest. For sixty feet of its length it was quite regular in form and had an average width of thirty inches and a depth of fourteen. The mud dug in making it was piled evenly along the lower side. Altogether it looked more like the work of a careful man with a shovel than of beaver without tools. Seepage and overflow from the ponds above filled and flowed slowly through it and out at the farther end, where it swept down the long slide into the big pond. Through this canal the logs had been taken one by one. At the farther end I found the butt-end log. It probably had been too heavy to heave out of the canal, but tracks in the mud indicated that there was a hard tussle before it was abandoned.

The pile of winter supplies was started. Close to the big house a few aspen leaves fluttered on twigs in the water; evidently these twigs were attached to limbs or larger pieces of aspen that

Harvest Time with Beavers

were piled beneath the surface. Could it be that the aspen which I had marked on the mountainside a quarter of a mile distant so short a time before, and which I had followed over slope and slide, through canal and basin, was now piled on the bottom of this pond? I waded out into the water, prodded about with a pole, and found several smaller logs. Dragging one of these to the surface, I found there were three notches in it.

Evidently these heavy green tree cuttings had been sunk to the bottom simply by the piling of other similar cuttings upon them. With this heavy material in the still water a slight contact with the bottom would prevent the drifting of accumulated cuttings until a heavy pile could be formed. However, in deep or swift water I have noticed that an anchorage for the first few pieces was secured by placing these upon the lower slope of the house or against the dam.

Scores of aspens were felled in the grove where the notched ones were. They were trimmed, cut into sections, and limbs, logs, and all taken over the route of the one I had followed, and at last placed in a pile beside the big house. This har-

In Beaver World

vest-gathering went on for a month. All about was busy, earnest preparation for winter. The squirrels from the tree-tops kept a rattling rain of cones on the leaf-strewn forest floor, the cheery chipmunk foraged and frolicked among the withered leaves and plants, while aspens with leaves of gold fell before the ivory sickles of the beaver. Splendid glimpses, grand views, I had of this strange harvest-home. How busy the beavers were! They were busy in the grove on the steep mountainside; they tugged logs across the runways; they hurried them across the water-basins, wrestled with them in canals, and merrily piled them by the rude house in the water. And I watched them through the changing hours; I saw their shadowy activity in the starry, silent night; I saw them hopefully leave home for the harvest groves in the serene twilight, and I watched them working busily in the light of the noonday sun.

Most of the aspens were cut off between thirteen and fifteen inches above the ground. A few stumps were less than five inches high, while a number were four feet high. These high cuttings

Harvest Time with Beavers

were probably made from reclining trunks of lodged aspens which were afterward removed. The average diameter of the aspens cut was four and one half inches at the top of the stump. Numerous seedlings of an inch diameter were cut, and the largest tree felled for this harvest measured fourteen inches across the stump. This had been laid low only a few hours before I found it, and a bushel of white chips and cuttings encircled the lifeless stump like a wreath. In falling, the top had become entangled in an alder thicket and lodged six feet from the ground. It remained in this position for several days and was apparently abandoned; but the last time I went to see it the alders which upheld it were being cut away. Although the alders were thick upon the ground, only those which had upheld the aspen had been cut. It may be that the beaver which felled them looked and thought before they went ahead with this cutting.

Why had this and several other large aspens been left uncut in a place where all were convenient for harvest? All other neighboring aspens were cut years ago. One explanation is that the beaver realized that the tops of the aspens were entangled and interlocked in the limbs of crowding spruces and would not fall if cut off at the bottom. This and one other aspen were the only large ones that were felled, and the tops of these had been recently released by the overturning of some spruces and the breaking of several branches on others. Other scattered large aspens were left uncut, but all of these were clasped in the arms of near-by spruces.

It was the habit of these colonists to transfer a tree to the harvest pile promptly after cutting it down. But one morning I found logs on slides and in canals, and unfinished work in the grove, as though everything had been suddenly dropped in the night when work was at its height. Coyotes had howled freely during the night, but this was not uncommon. In going over the grounds I found the explanation of this untidy work in a bear track and numerous wolf tracks, freshly moulded in the muddy places.

After the bulk of the harvest was gathered, I went one day to the opposite side of the moraine and briefly observed the methods of the Island

Harvest Time with Beavers

beaver colony. The ways of the two colonies were in some things very different. In the Spruce Tree Colony the custom was to move the felled aspen promptly to the harvest pile. In the Island Colony the custom was to cut down most of the harvest before transporting any of it to the pile beside the house. Of the one hundred and sixty-two trees that had been felled for this harvest, one hundred and twenty-seven were still lying where they fell. However, the work of transporting was getting under way; a few logs were in the pile beside the house, and numerous others were scattered along the canals, runways, and slides between the house and the harvest grove.

There was more wasted labor, too, in the Island Colony. This was noticeable in the attempts that had been made to fell limb-entangled trees that could not fall. One five-inch aspen had three times been cut off at the bottom. The third cut was more than three feet from the ground, and was made by a beaver working from the top of a fallen log. Still this high-cut aspen refused to come down and there it hung like a collapsed balloon entangled in tree-tops.

In Beaver World

Prowling hunters have compelled most beaver to work at night, but the Spruce Tree Colony was an isolated one, and occasionally its members worked and even played in the sunshine. Each day I secluded myself, kept still, and waited; and on a few occasions watched them as they worked in the light.

One windy day, just as I was unroping myself from the shaking limb of a spruce, I saw four beaver plodding along in single file beneath. They had come out of a hole between the roots of the spruce. At an aspen growth about fifty feet distant they separated. Though they had been closely assembled, each appeared utterly oblivious of the presence of the others. One squatted on the ground by an aspen, took a bite of bark out of it, and ate leisurely. By and by he rose, clasped the aspen with fore paws, and began to bite chips from it systematically. He was deliberately cutting it down. The most aged beaver waddled near an aspen, gazed into its top for a few seconds, then moved away about ten feet and started to fell a five-inch aspen. The one rejected was entangled at the top. Presently the third

Harvest Time with Beavers

beaver selected a tree, and after some trouble in getting comfortably seated, or squatted, also began cutting. The fourth beaver disappeared and I did not see him again. While I was looking for this one the huge, aged beaver whose venerable appearance had impressed me the first evening appeared on the scene. He came out of a hole beneath some spruces about a hundred feet distant. He looked neither to right nor to left, nor up nor down, as he ambled toward the aspen growth. When about halfway there he wheeled suddenly and took an uneasy survey of the open he had traversed, as though he had heard an enemy behind. Then with apparently stolid indifference he went on leisurely, and for a time paused among the cutters, which did nothing to indicate that they realized his presence. He ate some bark from a green limb on the ground, moved on, and went into the hole beneath me. He appeared so large that I afterward measured the distance between the two aspens where he paused. He was not less than three and a half feet long and probably weighed fifty pounds. He had all his toes; there was no white spot on his body; in fact, there was neither mark nor blemish by which I could positively identify him. Yet I feel that in my month around the colony I beheld the patriarch of the first evening in several scenes of action.

Sixty-seven minutes after the second beaver began cutting he made a brief pause; then he suddenly thudded the ground with his tail, hurriedly took out a few more chips, and ran away, with the other two beaver a little in advance, just as his four-inch aspen settled over and fell. All paused for a time close to the hole beneath me, and then the old beaver returned to his work. The one that had felled his tree followed closely and at once began on another aspen. The other beaver, with his aspen half cut off, went into the hole and did not again come out. By and by an old and a young beaver came out of the hole. The young one at once began cutting limbs off the recently felled aspen, while the other began work on the half-cut tree; but he ignored the work already done, and finally severed the trunk about four inches above the cut made by the other. Suddenly the old beaver whacked the ground and ran, but at thirty feet distant he paused and

Harvest Time with Beavers

nervously thumped the ground with his tail, as his aspen slowly settled and fell. Then he went into the hole beneath me.

This year's harvest was so much larger than usual that it may be the population of this colony had been increased by the arrival of emigrants from a persecuted colony down in the valley. The total harvest numbered four hundred and forty-three trees. These made a harvest pile four feet high and ninety feet in circumference. A thick covering of willows was placed on top of the harvest pile, — I cannot tell for what reason unless it was to sink all the aspens below reach of the ice. This bulk of stores together with numerous roots of willow and water plants, which are eaten in the water from the bottom of the pond, would support a numerous beaver population through the days of ice and snow.

When I took my last tour through the colony everything was ready for the long and cold winter. Dams were in repair and ponds were brimming over with water, the fresh coats of mud on the houses were freezing to defy enemies, and a bountiful harvest was home. Harvest-gathering is full

In Beaver World

of hope and romance. What a joy it must be to every man or animal who has a hand in it! What a satisfaction, too, for all dependent upon a harvest, to know that there is abundance stored for all the frosty days!

The people of this wild, strange, picturesque colony had planned and prepared well. I wished them a winter unvisited by cruel fate or foe, and trusted that when June came again the fat and furry young beavers would play with the aged one amid the tiger lilies in the shadows of the big spruce trees.

Transportation Facilities

Transportation Facilities

Two successive dry years had greatly reduced the water-level of Lily Lake, and the consequent shallowness of the water made a serious situation for its beaver inhabitants. This lake covered about ten acres, and was four feet deep in the deepest part, while over nine tenths of the area the water was two feet or less in depth. It was supplied by springs. Early in the autumn of 1911 the water completely disappeared from about one half of the area, and most of the remainder became so shallow that beaver could no longer swim beneath the surface. This condition exposed them to the attack of enemies and made the transportation of supplies to the house slow and difficult.

In the lake the beaver had dug an extensive system of deep canals, — the work of years. By means of these deep canals the beaver were able to use the place until the last, for these were full of water even after the lake-bed was completely exposed. One day in October while passing the lake, I noticed a coyote on the farther shore stop suddenly, prick up his ears, and give alert attention to an agitated forward movement in the shallow water of a canal. Then he plunged into the water and endeavored to seize a beaver that was struggling forward through water that was too shallow for his heavy body. Although this beaver made his escape, other members of the colony may not have been so fortunate.

The drouth continued and by mid-October the lake went entirely dry except in the canals. Off in one corner stood the beaver house, a tiny rounded and solitary hill in the miniature black plain of lake-bed. With one exception the beaver abandoned the site and moved on to other scenes, I know not where. One old beaver remained. Whether he did this through the fear of not being equal to the overland journey across the dry rocky ridge and down into Wind River, or whether from deep love of the old home associations, no one can say. But he remained and endeavored to make provision for the oncoming winter. Close to the house he dug or enlarged

LAKE-BED CANALS AT LILY LAKE, OCTOBER, 1911

SECTION OF A 750-FOOT CANAL AT LILY LAKE
Here five feet wide and three feet deep

Transportation Facilities

a well that was about six feet in diameter and four feet in depth. Seepage filled this hole, and into it he piled a number of green aspen chunks and cuttings, a meagre food-supply for the long, cold winter that followed. Extreme cold began in early November, and not until April was there a thaw.

Before the lake-bed was snow-covered, all the numerous canals and basins which the beaver had excavated could be plainly seen and examined. The magnitude of the work which the beaver had performed in making these is beyond comprehension. I took a series of photographs of these excavations and made numerous measurements. To the north of the house a pool had been dug that was three feet deep, thirty feet long, and about twenty wide. There extended from this a canal that was one hundred and fifty feet long. The food basin was thirty feet wide and four feet deep. This had a canal connection with the house. In the bottom of the basin was one of the feeble springs which supply the lake. Another canal, which extended three hundred and fifty feet in a northerly direction from the house, was from three to four feet wide and three feet deep. The largest ditch or canal was seven hundred and fifty feet long and three feet deep throughout. This extended eastward, then northeasterly, and for one hundred feet was five feet wide. In the remaining six hundred and fifty feet it was three to four feet wide. There were a number of minor ditches and canals connecting the larger ones, and altogether the extent of all made an impressive show in the empty lake-basin.

Meantime the old beaver had a hard winter. The cold weather persisted, and finally the well in which he had deposited winter food froze to the bottom. Even the entrance-holes into the house were frozen shut. This sealed him in. The old fellow, whose teeth were worn and whose claws were bad, apparently tried in vain to break out. On returning from three months' absence, two friends and I investigated the old beaver's condition. We broke through the frozen walls of the house and crawled in. The old fellow was still alive, though greatly emaciated. For some time — I know not how long — he had subsisted on the wood and the bark of some green sticks

Transportation Facilities

which had been built into an addition of the house during the autumn. We cut several green aspens into short lengths and threw them into the house. The broken hole was then closed. The old fellow accepted these cheerfully. For six weeks aspens were occasionally thrown to him, and at the end of this time the spring warmth had melted the deep snow. The water rose and filled the pond and unsealed the entrance to the house, and again the old fellow emerged into the water. The following summer he was joined, or rejoined, by a number of other beaver.

In many localities the canals or ditches dug and used by the beaver form their most necessary and extensive works. These canals require enormous labor and much skill. In point of interest they even excel the house and the dam. It is remarkable that of the thousands of stories concerning the beaver only a few have mentioned the beaver canals. These are labor-saving improvements, and not only enable the beaver to live easily and safely in places where he otherwise could not live at all, but apparently they allow him to live happily. The excavations made in

In Beaver World

taking material for house or dam commonly are turned to useful purpose. The beaver not only builds his mound-like house, but uses the basin thus formed in excavating earthy material for the house for a winter food depository. Ofttimes, too, in building the dam he does it by piling up the material dug from a ditch which runs parallel and close to the dam, and which is useful to him as a deep waterway after the dam is completed.

In transporting trees for food-supply, water transportation is so much easier and safer than land, that wherever the immediate surroundings of the pond are comparatively level the beaver endeavors to lead water out to tree groves by digging a canal from the edge of the pond to these groves. The felled trees are by this means easily floated into the pond. One of the simplest forms of beaver canal is a narrow, outward extension of the pond. This varies in length from a few yards to one hundred feet or more.

Another and fairly common form of canal is one that is built across low narrow necks of land which thrust out into large beaver ponds, or on

Transportation Facilities

narrow stretches of land around which crooked streams wander.

The majority of beaver ponds are comparatively shallow over the greater portion of their area. In many cases it is not easy, or even possible, to deepen them. They may be so shallow that the pond freezes to the bottom in winter except in its small deeper portion. The shallow ponds are made more usable by a number of canals in the bottom. These canals assure deep-water stretches under all conditions. Most beaver ponds have a canal that closely parallels the dam. In some instances this is extended around the pond a few yards inside the shore-line. Two canals usually extend from the house. One of these connects with the canal by the dam, the other runs to the place on the shore (commonly at the end of a trail or slide) most visited by the beaver.

In Jefferson Valley, Montana, not far from Three Forks, I enjoyed the examination of numerous beaver workings, and made measurements of the most interesting system of beaver canals that I have ever seen. The beaver house for which these canals did service was situated on the south bank of the river, about three feet above the summer level of the water and about two hundred feet north of the hilly edge of the valley. From the river a crescent-shaped canal, about thirty-five feet in length, had been dug halfway around the base of the house. Connected with this was a basin for winter food; this was five feet deep and thirty-five feet in diameter. From this a canal extended southward two hundred and seven feet. One hundred and ten feet distant from the house was a boulder that was about ten feet in diameter. This was imbedded in about two feet of soil. Around this boulder the canal made a detour, and then resumed its comparatively straight line southward.

Over the greater portion of its length this canal was four feet wide, and at no point was it narrower than three feet. Its average depth was twenty-eight inches. For one hundred and forty-seven feet it ran through an approximately level stretch of the valley, and seepage filled it with water. A low, semi-circular dam, about fifty feet in length, crossed it at the one-hundred-and forty-seven-foot mark, and served to catch and

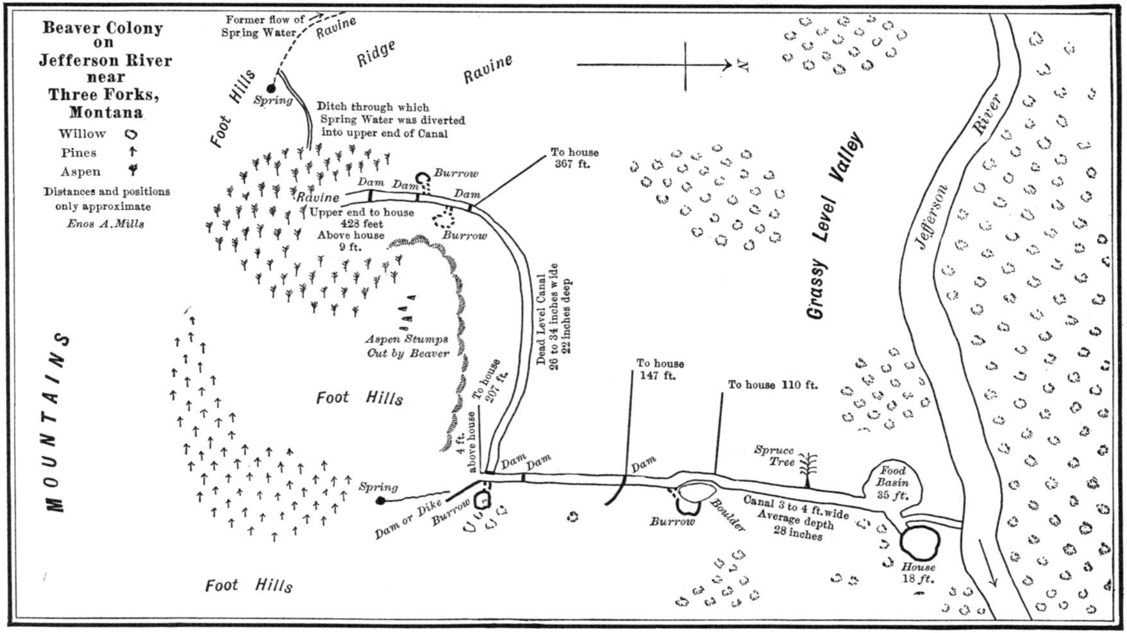
Beaver Colony
on
Jefferson River
near
Three Forks,
Montana
Willow
Pines
Aspen
Distances and positions
only approximate
Enos A. Mills
Former flow of
Spring Water
Ravine
Ridge
Ravine
N
Foot Hills
Spring
Ditch through which
Spring Water was diverted
into upper end of Canal
To house
367 ft.
Burrow
Dam
Dam
Dam
Ravine
Upper end to house
428 feet
Above house
9 ft.
Burrow
Dead Level Canal
26 to 34 inches wide
22 inches deep
Grassy Level Valley
Jefferson
River
Aspen Stumps
Cut by Beaver
MOUNTAINS
Foot Hills
To house
207 ft.
4 ft.
above house
To house
147 ft.
To house 110 ft.
Spruce
Tree
Food
Basin
35 ft.
Dam
Dam
Dam
Spring
Dam or Dike
Burrow
Burrow
Boulder
Canal 3 to 4 ft. wide
Average depth
28 inches
House
18 ft.
Foot Hills

Transportation Facilities

run seepage water into it, and also to act as a wall across the canal to hold the water. The most southerly sixty feet of this canal on the edge of the foothills ran uphill, and was about four feet deep at the upper end, four feet higher than the end by the house. The dam across it was supplemented by a wall forty-eight feet further on. This wall was simply a short dam across the canal, in a part that was inclined, and plainly for the purpose of retaining water in the canal. The upper part of the canal was filled with water by a streamlet from off the slope. Apparently this canal was old, for there was growing on its banks near the house, a spruce tree, four inches in diameter, that had grown since the canal was made.

The wall or small dam which beaver build across canals that are inclined represents an interesting phase of beaver development. That these walls are built for the purpose of retaining water in the canal appears certain. They are most numerous in canals of steepest incline, though rarely less than twenty feet apart. I have not seen a wall in an almost dead-level canal, except it was there for the purpose of raising the height of the water. This wall or buttress is after all but a dam, and like most dams it is built for the purpose of raising and maintaining the level of water.

Extending at right angles westward from the end of the old canal was a newer one of two hundred and twenty-one feet. A wall separated and united the two. One hundred and sixty feet of this new canal ran along the contour of a hill, approximately at a dead level. Then came a wall, and from this the last sixty-one feet extended southward up a shallow ravine. In this part there were two walls. The upper end of the sixty-one-foot extension was nine feet higher than the house, and four hundred and twenty-eight feet distant from it. The two-hundred-and-twenty-one-foot extension was from twenty-six to thirty-four inches wide, and averaged twenty-two inches deep. The entire new part was supplied with spring water, which the beaver had diverted from a ravine to the west and led by a seventy-foot ditch into the upper end of their canal. Thirty feet from the end of the canal were two burrows, evidently safe places into which the beaver could

Transportation Facilities

retreat in case of sudden attack from wolves or other foe. There were two other of these burrows, one at the outer end of the old canal and the other alongside the boulder one hundred and ten feet from the house.

At the time I saw these canals, the only trees near were those of an aspen grove which surrounded the extreme end. It was autumn, and on both tributary slopes by the end of the canal, aspens were being cut, dragged, and rolled down these slopes into the upper end of the canal, then floated through its waters, dragged over and across the walls, and at last piled up for winter food in the basin by the house. In all probability this long, large canal had been built a few yards at a time, being extended as the trees near-by were cut down and used.

Where beaver long inhabit a locality it is not uncommon for them to have two or three distinct and well-used trails from points on the water's edge which lead into neighboring groves or tree-clumps. These are the beaten tracks traveled by the beaver as they go forth from the water for food, and over which they drag their trees and saplings into the water. On steep slopes by the water these are called slides. This name is also given to places in the dam over which beaver frequently pass in their outgoings and incomings. Commonly these trails avoid ridges and ground swells by keeping in the bottom of a ravine; logs are cut through and rolled out of the way, or a tunnel driven beneath; obstructions are removed, or a good way made round them. Their log roads compare favorably with the log roads of woodsmen who cut with steel instead of enamel.

In most old beaver colonies, where the character of the bottom of the pond permits it, there are two or more tunnels or subways beneath the floor of the principal pond. The main tunnel begins close to the foundation of the house, and penetrates the earth a foot or more beneath the water to a point on land a few feet beyond the shore-line. If there are a number of small ponds in a colony that are separated by fingers of land, it is not uncommon for these bits of land to be penetrated by a thoroughfare tunnel. These tunnels through the separating bits of land enable the beaver to go from one pond to another with-

Transportation Facilities

out exposing themselves to dangers on land, and also offer an easy means of intercommunication between ponds when these are ice-covered. Pond subways also afford a place of refuge or a means of escape in case the house is destroyed, the dam broken, or the pond drained, or in case the pond should freeze to the bottom. Commonly these are full of water, but some are empty. On the Missouri and other rivers, where there are several feet of cut banks above the water, beaver commonly dug a steeply inclined tunnel from the river's edge to the top of a bank a few feet back.

Most of this tunnel work is hidden and remains unknown. A striking example was in the Spruce Tree Colony, elsewhere described. These colonists, apparently disgusted by having their ponds completely filled with sediment which came down as the result of a cloudburst, abandoned the old colony-site. A new site was selected on a moraine, only a short distance from the old one. Here in the sod a basin was scooped out, and a dam made with the excavated material. The waters from a spring which burst forth in the moraine, about two hundred yards up the slope

In Beaver World

and perhaps one hundred feet above, trickled down and in due time formed a pond. The following year this pond was enlarged, and another one built upon a terrace about one hundred feet up the slope. From year to year there were enlargements of the old pond and the building of new pondlets, until there were seven on the terraces of this moraine. These, together with the connecting slides and canals, required more water than the spring supplied, especially in the autumn when the beaver were floating their winter supplies from pond to pond. Within the colony area, too, were many water-filled underground passages or subway tunnels. One of these penetrated the turf beneath the willows for more than two hundred feet.

While watching the autumnal activities of this colony, as described in another chapter, I broke through the surface and plunged my leg into an underground channel or subway that was half filled with water. Taking pains to trace this stream downward, I found that it emptied into the uppermost of the ponds along with the waters from a small spring. Then, tracing the channel

Transportation Facilities

upwards, I found that, about one hundred and forty feet distant from the uppermost pond, it connected with the waters of the brook on which the old colony formerly had a place. This tunnel over most of its course was about two feet beneath the surface, was fourteen inches in diameter, and ran beneath the roots of spruce trees. The water which the tunnel led from the brook plainly was being used to increase the supply needed in the canals, ponds, and pools of the Spruce Tree Colony. The intake of this was in a tiny pond which the beaver had formed by a damlet across the brook. That this increased supply of water was of great advantage to the busy and populous Spruce Tree Colony, there can be no doubt. Was this tunnel planned and made for this especial purpose, or was the increased water-supply of the colony the result of accident by the brook's breaking into this subway tunnel?

The canals which beaver dig, the slides which they use, the trails which they clear and establish, conclusively show that these animals appreciate the importance of good waterways and good roads, — in other words, good transportation facilities.

The Primitive House

The Primitive House

THE LILY LAKE beaver house, in which the old beaver spent the drouthy winter, was a large roughly rounded affair that measured twenty-two feet in diameter. It rose only four feet above the normal water-line. This house had been three times altered and enlarged, and once raised in height. Its mud walls were heavily reinforced with polelike sticks, which were placed at the junctures of the enlargements. The one large room was more than twelve feet in diameter. Near the centre stood a support for the upper part of the house. This support was about one and a half by two and a half feet, and was composed for the most part of sticks. But few houses have this support; commonly the room is vaulted. The room itself averaged two and a half feet high. It had four entrances.

A house commonly has two entrances, but it may have only one or as many as five. Thus the way to the outer world from the inside of the house is through one or more inclined passageways or tunnels. The upper opening of these entrances is in the floor a few inches above the water-level, and the lower opening in the bottom of the pond under about three feet of water. These extend at an angle through the solid foundation of the house, are about one foot in diameter and four to fifteen feet long, and are full of water almost to floor-level. This dark, windowless hut has no other entrance.

Most beaver houses stand in a pond, though a number are built on the shore and partly in the water, and still others on the bank a few feet away from the water. The external appearance and internal construction of the houses are in a general way the same, regardless of the situation or size. Most beaver houses appear conical. Measured on the water-line, they are commonly found to be slightly elliptical. The diameter on the water-line is from five to thirty-five feet, and the height above water is from three to seven feet.

A house may be built almost entirely of sticks, or of a few sticks with a larger proportion of mud and turf. In building, a small opening is left, —

The Primitive House

or built around and over, — which is afterwards enlarged into a room.

Houses that are built in a pond usually stand in three or four feet of water. The foundation is laid on the bottom of the pond, of the size intended for the house, and built up a solid mass to a few inches above water-level. This island-like foundation is covered with a crude hemisphere or dome-shaped house, the central portion of the foundation forming the floor of the low-vaulted room which is enclosed by the thick house-walls. In building the house the beaver provide a temporary support for the combined roof and walls by piling in the centre of the floor a two-foot mound of mud. Over this is placed a somewhat flattened tepee- or cone-shaped frame of sticks and small poles. These stand on the outer part of the foundation and lean inward with upper ends meeting against and above the temporary support. The beaver then cover this framework with two or three feet of mud, brush, and turf, and thus make the walls and the roof of the house. When the outer part of the house is completed, they dig an inclined passageway,

In Beaver World

from the bottom of the pond up through the foundation, into the irregular space left between the supporting pile of mud and the walls. And of this space they shape a room, by clawing out the temporary support and gnawing off the intruding sticks. This represents the most highly developed type of beaver house.

In most houses the temporary support is not used, but a part of the wall is carried up to completion, and against it are leaned sticks, which rest upon the edge of the remaining foundation. A finished house of this kind has a slightly elliptical outline. However, many a house is a crude haphazard pile of material in which a room has been burrowed.

The room is from one to three feet high, and from three to twenty feet across. The room is a kind of a burrow and is without either door or window. Half-buried sticks make a comparatively dry floor, despite the fact that it is only a few inches above water-level. Beaver sleep on the floor, usually with tail bent along the side after the fashion of a dozing cat, in a nest of shredded wood, which they patiently make by thinly split-

The Primitive House

ting and paring pieces of wood. Just why this kind of bedding is used cannot be said, but probably because this material dries more quickly, is more comfortable and more sanitary, and harbors fewer parasites. However, a few beds are made of grass, leaves, or moss.

But little earthy matter is used in the tip-top of the house, where the minute disjointed air-holes between the interlaced poles give the room scanty ventilation.

Except in a few cases where house-walls are overgrown with willows or grass, the erosive action of wind and water rapidly thins and weakens them. Hence the house must receive frequent repairs. Each autumn it is plastered or piled all over with sticks or mud. The mud covering varies in thickness from two to six inches. The mud for this purpose is usually dredged from the bottom of the pond close to the foundation of the house. It is carried up, a double handful at a time, the beaver waddling on his hind legs as he holds it with his fore paws against his breast. A half-dozen or more beaver may be carrying mud up at once. The covering not only thickens the walls and increases the warmth of the house, but also freezes and becomes an armor of stone that is impregnable to most beaver enemies. The "mudding" of the house is a part of the natural and necessary preparation for winter. It may also be a special means of protection deliberately carried out by the beaver. The fact that an occasional thick-walled or grass-covered beaver house was not thus plastered in autumn — perhaps because it did not need it — has led a few people to affirm that beaver houses are not mud-covered in the autumn. Many years of observation show that most beaver houses do receive an autumnal plastering, and the few that do not have this attention usually have thick, well-preserved walls and do not need it.

One autumn in Montana, of twenty-seven beaver houses which I examined, twenty-one received mud covering; three of the others were thickly overgrown with willows and two were grass-grown. Only one thin-walled house that needed reinforcement did not receive it; and this one, by the way, was broken into by a bear before the winter had got fairly under way.

AN UNPLASTERED AND A PLASTERED HOUSE

The Primitive House

In the autumn of 1910 I made notes concerning eighteen houses. These I watched during October and November. Thirteen were plastered; a willow-grown one and a weed-grown one, both of which had thick walls, were not plastered. The remaining three were not greatly in need of additional thickness, so received only a scanty covering of sticks. Two of these were broken into by some animal during the winter, while none of the others were disturbed.

Beaver frequently show good judgment in that important matter of selecting a site for the house. Ice and sediment are two factors with which the beaver must constantly contend. In the pond the house is commonly placed in deep water, and apparently where the depth around it will not be rapidly reduced by the depositing of sediment. Keeping the house-entrance, the harvest-pile basin, and the canals from filling with sediment is one of the difficult problems of beaver life.

To guard against the rapid encroachments of the deposits of sediment, one group of beaver, apparently with forethought, built a dam that formed a pond from the waters of a small spring which carried but little or no sediment. I have noticed a number of instances in which a pond was made on a small streamlet with greater labor than it would have required to form a pond in a near-by brook. As there were a number of other conditions favorable to the brook situation of the house, the only conclusion I could reach was that these selections for colony-sites were made with the intention of avoiding the ever-encroaching sediment, — for in some beaver ponds this sediment is deposited annually to the depth of several inches.

Ice is one of the troubles of beaver existence. It is of the utmost importance to the beaver that he should have his house so situated that the ice of winter does not close the entrance to it, and also that the deep water in which his pile of green provisions is deposited does not freeze solid and thus exclude him from the food-supply. The ice fills the pond from the top and compels him to be constantly vigilant to save himself from its encroachments. Many a beaver home has been built alongside a spring, around which the beaver dredged a deep hole and in this

The Primitive House

deposited the winter food-supply. The constant flow of the spring water prevented thick ice from forming, both around the food-pile and between it and the house-entrance.

Large numbers of beaver do not possess a house. Beaver who live without a dam or pond commonly do not build a house, but are content with a burrow or a number of burrows in the banks of the waters which they inhabit. In the severe struggle to live, there is a tendency on the part of the beaver to avoid the building of dams and houses, as these reveal their presence and put the aggressive trapper on their trail.

Many colonies have both houses and burrows. Apparently the houses were used in the wintertime, the burrows in summer. One beaver burrow which I examined was about one foot above the level of the pond and twelve feet distant from it. The entrance tunnels were sixteen feet in length, and began a trifle more than three feet under water near the edge of the pond. This burrow measured five and a half feet long, about half as wide, and seventeen inches high. It was immediately beneath the outspreading roots of an Engelmann spruce. The majority of beaver burrows are about two thirds the size of this one.

One November I examined more than a score of beaver colonies. There was no snow, but recent cold had covered the pond with ice and solidified the miry surroundings. Over the frozen surface I moved easily about and made many measurements. One of these colonies was a fairly typical one. The colony was on a swift-running stream that came down from the snowy heights, three miles distant. The top of Long's Peak and Mt. Meeker looked down upon the scene. The altitude of this colony was about nine thousand feet. The ponds were in part surrounded by semi-boggy willow flats, with here and there a high point or a stretch of bank that was covered with aspens. The tops of a few huge boulders thrust up through the water. All around stood guard a tall, dark forest of lodge-pole pines. These swept up the mountainside, where they were displaced by a growth of Engelmann spruce which reached up to timber-line on the heights above.

This colony had a number of ponds, with a few short canals extending outward from them.

The Primitive House

A conical house of mud and slender poles stood in the larger pond. Above this pond there were half a dozen pondlets, the uppermost of which was formed across the brook by a semi-circular dam. Over the outward ends of this dam the water flowed and was caught in other ponds; these in turn overflowed, the water traversing two other ponds, one below the other, just above the main one. Below the large pond were three smaller ones in close succession. The dam of each pond backed the water against the dam above it.

The dam of the main pond was two hundred and thirty feet long. Each end bent upward at a sharp angle and extended a number of yards upstream. This dam measured five feet at its highest point, but along the greater portion was only a trifle more than three feet high. The central part was overgrown with sedge and willows and appeared old; but the extreme ends appeared new, and probably had been in part constructed within a few weeks. The whole dam was formed of earth and slender poles. The pond formed by it was one hundred and eighty feet wide, and had an average length up and down stream of one hundred and ten feet. The average depth was only two feet.

Near the centre of this large pond stood the house, a trifle nearer to the dam than to the upper edge of the pond. I measured it on the water- or rather the ice-level. It took twenty-six feet of rope to go around it. The top of the house rose exactly five feet above the ice. The house was built of a mixture of sods and willow sticks. The ends of the sticks here and there thrust out through the three-or-four-inch covering of mud which the house had recently received. Wondering how much of the house was in the water below the level of the ice, I thought to measure the depth by thrusting a pole through the ice to the bottom. Holding it in an upright position, I raised it and brought it down with all my strength. The pole went through the ice and so did I. The water was three feet deep. This depth covered only a small area around the house and was maintained by frequent digging. The house is often plastered with this dredged material. Altogether, then, the house from its lowest founda-

The Primitive House

tion on the bottom of the pond to the conical top was eight feet high. The foundation of this house was made of turf, masses of grass roots, and a small percentage of mud thickly reinforced with numerous willow sticks. The floor was mostly sticks. As the entrance tunnels were filled with water to a point about three inches below the floor-level, and as these were the only entrances or openings into the house, friend or foe could enter only by coming up through one or the other of these water-filled tunnels from the bottom of the pond.

The single, circular, dome-like room of this house was four and a half feet in diameter and about two feet in height. Its ceiling was roughly formed by a confused interlacing of sticks, which stood at an angle. The spaces between were filled with root-matted mud. The walls were a trifle more than two feet thick, except around the conical top. Here was a small space, mostly of interlacing sticks, the thickness of which was but one foot. As very little mud had been used in this part, there were thus left a few tiny air-holes. As I approached, there could be seen arising from these holes the steamy and scented breath of the beaver inhabitants within. Since the ventilation of beaver houses is exceedingly poor, and as this animal probably does not suffer from tuberculosis, it is possible that ventilation is assisted, and some of the impure air absorbed by the water, which rises almost to the floor in the large entrance-holes.

The early trappers from time to time noted extended general movements or emigrations among beaver, which embraced an enormous area. They, as with human emigrants, probably were seeking a safer, better home. Some of these movements were upstream, others down; commonly away from civilization, but occasionally toward it. For this the Missouri River was the great highway. Limited emigrations of this kind still occasionally occur.

The annual migration is a different affair. This has been noted for some hundred and fifty years or more, and probably has gone on for centuries. This peculiar migration might be called a migratory outing. In it all members of the colony appear to have taken part, leaving home in June,

The Primitive House

scattering as the season advanced. Rambles were made up and down stream, other beaver settlements visited, brief stays made at lakes, adventures had up shallow brooks, and daring journeys made on portages. The country was explored. The dangers and restrictions imposed during the last twenty-five years appear in some localities to have checked this movement, and in others to have stopped it completely. But in most colonies it still goes on, though probably not usually enjoyed by mothers and children except to a limited extent.

By the first of September all have returned to the home, or joined another colony or assembled at the place where a new colony is to be founded. This annual vacation probably sustained the health of the colonists; they got away from the parasites and the bad air of their houses. The outing was taken for the sheer joy of it. Incidentally, it brought beaver into new territory and acquainted them with desirable colony-sites and the route thereto, — useful information in case the colonists were compelled suddenly to abandon the old home.

It is natural for the beaver to be silent. In

In Beaver World

silence he becomes intimate with the elements, and, while listening, hears and understands all moods and movements that concern him. He is a master in translating sound. It wakens or warns, threatens or gladdens, and woos him back to slumberland.

On the wild frontier in his fortress island home in safety he sits and sleeps in darkness. He cannot see outside, but the ever-changing conditions of the surrounding outer world are revealed to him by continuous and varying sounds that penetrate the thick windowless walls of his house. He hears the cries of the coyote and the cougar, the call of moose, the wild and fleeting laugh of the kingfisher, the elemental melody of the ouzel, and many an echo faintly from afar. He hears the soft vibrations from the muffled feet of enemies; and, above his head, the raking threat of claws upon the top of his house. Endlessly the water slides and gently pours over the dam, and softly ebbs around the pond's primeval shores. The earthquake thunder warns of storm, the floods roar; then through day and night the cleared and calmed stream goes by. The wind booms among

The Primitive House

the baffling pines, and the broken and leafless tree falls with a crash! There is silence! Along the stream's open way through the woods numberless breezes whisper and pause by the primitive house in the water.

The Beaver's Engineering

The Beaver's Engineering

REALIZING that the supply of aspens near the waters of the Moraine Colony close to my home was almost exhausted, I wondered whether it would be possible for the beavers to procure a sufficient supply downstream, or whether they would deem it best to abandon this old colony and migrate.

Out on the plains, where cottonwoods were scarce, the beavers first cut those close to the colony, then harvested those upstream, sometimes going a mile for them, then those downstream; but rarely were the latter brought more than a quarter of a mile. If enemies did not keep down the population of a colony so situated, it was only a question of time until the scarcity of the food-supply compelled the colonists to move either up or down stream and start anew in a place where food trees could be obtained. But not a move until necessity drove them!

Not far from my home in the mountains the

In Beaver World

inhabitants of two old beaver colonies endured hardships in order to remain in the old place. One colony, in order to reach a grove of aspens, dug a canal three hundred and thirty-four feet long, which had an average depth of fifteen inches and a width of twenty-six inches. It ended in a grove of aspens, which were in due time cut down and floated through this canal into the pond, alongside the beaver house. The other colony endured dangers and greater hardships.

During the summer of 1900 an extensive forest fire on the northerly slope of Long's Peak wrought great hardship among beaver colonies along the streams in the fire district. This fire destroyed all the aspens and some of the willows. In order to have food while a new growth of aspens was developing, the beavers at a colony on the Bierstadt Moraine were compelled to bring their winter supply of aspens the distance of a quarter of a mile from an isolated grove that had escaped the fire. This stood on a bench of the moraine at an altitude about fifty feet greater than that of the beaver pond. Aspens from the grove were dragged about two hundred feet, then floated

THE 334-FOOT CANAL

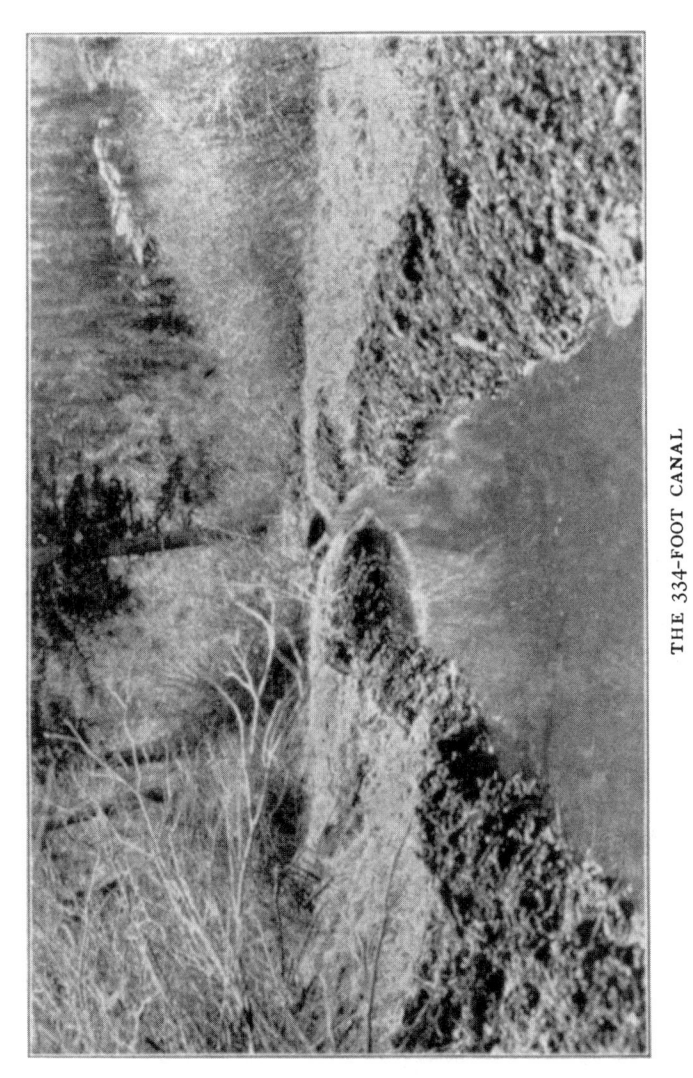

The Beaver's Engineering

across a small water-hole, and from this taken up the steep slope of a ridge, then down to a point about one hundred feet from the pond. Between this place and the pond was a deep wreckage of fire-killed and fallen spruces. To cut an avenue through these was too great a task for the beavers; so with much labor they dug a canal beneath the wide heap of wreckage, and through this, beneath the gigantic fallen trees, the harvested aspens were dragged and piled in the pond for winter food. The gathering of these harvests, even by beavers, must have been almost a hopeless task. In going thus far from water many of the harvesters were exposed to their enemies, and it is probable that many beavers lost their lives.

Beavers become strongly attached to localities and especially to their homes. It is difficult to drive them away from these, but the exhaustion of the food-supply sometimes compels an entire colony to abandon the old home-site, migrate, and found a new colony. Some of the beavers' most audacious engineering works are undertaken for the purpose of maintaining the food-supply of the colony. It occasionally happens that the food trees near the water by an old colony become scarce through excessive cutting, fires, or tree diseases. In cases of this kind the colonists must go a long distance for their supplies, or move. They prefer to stay at the old place, and will work for weeks and brave dangers to be able to do this. They will build a dam, dig a new canal, clear a difficult right-of-way to a grove of food saplings, and then drag the harvest a long distance to the water; and now and then do all these for just one more harvest, one more year in the old home.

The Moraine Colony had lost its former greatness. Instead of the several ponds and the eight houses of which it had consisted twenty years before, only one house and a single pond remained. The house was in the deep water of the pond, about twenty feet above the dam. A vigorous brook from Chasm Lake, three thousand feet above, ran through the pond and poured over the dam near the house. The colony was on a delta tongue of a moraine. Here it had been established for generations. It was embowered in a young pine forest and had ragged areas of willows

The Beaver's Engineering

around it. A fire and excessive cutting by beavers had left but few aspens near the water. These could furnish food for no more than two autumn harvests, and perhaps for only one. Other colonies had met similar conditions. How would the Moraine Colony handle theirs?

The Moraine colonists mastered the situation in their place with the most audacious piece of work I have ever known beavers to plan and accomplish. About one hundred and thirty feet south of the old pond was a grove of aspens. Between these and the pond was a small bouldery flat that had a scattering of dead and standing spruces and young lodge-pole pines. A number of fallen spruces lay broken among the partly exposed boulders of the flat. One day I was astonished to find that a dam was being built across this flat, and still more astonished to discover that this dam was being made of heavy sections of fire-killed trees. Under necessity only will beavers gnaw dead wood, and then only to a limited extent. Such had been my observations for years; but here they were cutting dead, fire-hardened logs in a wholesale manner. Why were they cutting this dead wood, and why a dam across a rocky flat,—a place across which water never flowed? A dam of dead timber across a dry flat appeared to be a marked combination of animal stupidity,—but the beavers knew what they were doing. After watching their activities and the progress of the dam daily for a month, I realized that they were doing development work, with the intention of procuring a food-supply. They completed a dam of dead timber.

At least two accidents happened to the builders of this dead-wood dam. One of these occurred when a tree which the beavers had gnawed off pinned the beaver that had cut it between its end and another tree immediately behind the animal. The other accident was caused by a tree falling in an unexpected direction. This tree was leaning against a fallen one that was held several feet above the earth by a boulder. When cut off, instead of falling directly to the earth it slid alongside the log against which it had been leaning and was shunted off to one side, falling upon and instantly killing two of the logging beavers.

The dam, when completed, was eighty-five feet

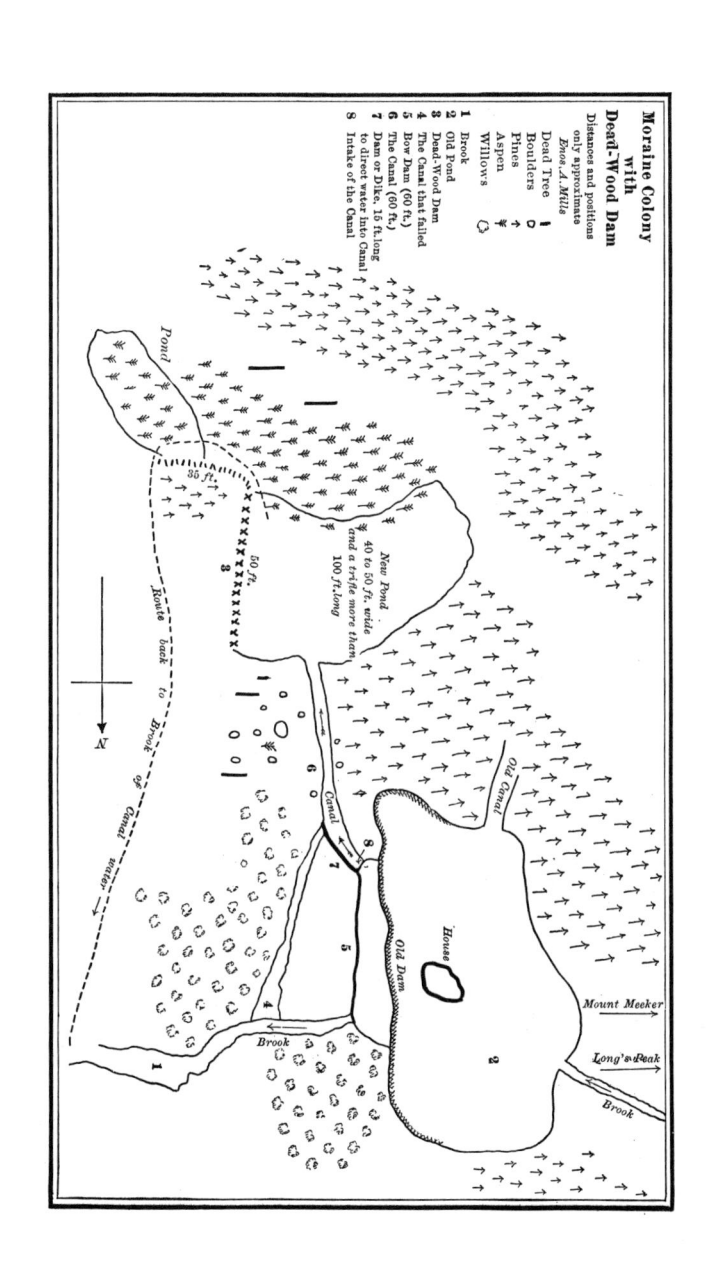

Moraine Colony with Dead-Wood Dam

Distances and positions only approximate

Enos A. Mills

Dead Tree	❘
Boulders	▢
Pines	↑
Aspen	✳
Willows	◯

1. Brook
2. Old Pond
3. Dead-Wood Dam
4. The Canal that failed
5. Bow Dam (60 ft.)
6. The Canal (60 ft.)
7. Dam or Dike, 16 ft. long to direct water into Canal
8. Intake of the Canal

The Beaver's Engineering

long. It was about fifty feet below the main pond and sixty feet distant from the south side of it. Fifty feet of the new dam ran north and south, parallel to the old one; then, forming a right angle, it extended thirty-five feet toward the east. It averaged three feet in height, being made almost entirely of large chunks, dead-tree cuttings from six to fifteen inches in diameter and from two to twelve feet long. It appeared a crude windrow of dead-timber wreckage.

The day it was completed the builders shifted the scene of activity to the brook, a short distance below the point where it emerged from the main pond. Here they placed a small dam across it and commenced work on a canal, through which they endeavored to lead a part of the waters of the brook into the reservoir which their deadwood dam had formed.

There was a swell or slight rise in the earth of about eighteen inches between the reservoir and the head of the canal that was to carry water into it. The swell, I suppose, was not considered by the beavers. At any rate, they completed about half the length of the canal, then appar-

In Beaver World

ently discovered that water would not flow through it in the direction desired. Other canal-builders have made similar errors. The beavers were almost human. This part of the canal was abandoned and a new start made. The beavers now apparently tried to overcome the swell in the earth by an artificial work.

A pondlet was formed immediately below the old pond by building a sixty-foot bow-like dam, the ends of which were attached to the old dam. The brook pouring from the old pond quickly filled this new narrow, sixty-foot-long reservoir. The outlet of this was made over the bow dam at the point nearest to the waiting reservoir of the dead-wood dam. The water, where it poured over the outlet of the bow dam, failed to flow toward the waiting reservoir, but was shed off to one side by the earth-swell before it. Instead of flowing southward, it flowed eastward. The beavers remedied this and directed the flow by building a wing dam, which extended southward from the bow dam at the point where the water over-poured. This earthwork was about fifteen feet long, four feet wide, and two high. Along the

The Beaver's Engineering

upper side of this the water flowed, and from its end a canal was dug to the reservoir.

About half of the brook was diverted, and this amount of water covered the flat and formed a pond to the height of the dead-wood dam in less than three days. Most of the leaky openings in this dam early became clogged with leaves, trash, and sediment that were carried in by the water, but here and there were large openings which the beavers mudded themselves. The new pond was a little more than one hundred feet long and from forty to fifty feet wide. Its southerly shore flooded into the edge of the aspen grove which the beavers were planning to harvest.

The canal was from four to five feet wide and from eight to twenty inches deep. The actual distance that lay between the brook and the shore of the new pond was ninety feet. Though the diverting of the water was a task, it required less labor than the building of the dam.

With dead timber and the canal, the beavers had labored two seasons for the purpose of getting more supplies without abandoning the colony. If in building the dam they had used the green,

In Beaver World

easily cut aspens, they would have greatly reduced the available food-supply. It would have required most of these aspens to build the dam. The only conclusion I can reach is that the beavers not only had the forethought to begin work to obtain a food-supply that would be needed two years after, but also, at the expense of much labor, actually saved the scanty near-by food-supply of aspens by making their dam with the hard, fire-killed trees.

A large harvest of aspen and willow was gathered for winter. Daily visits to the scene of the harvest enabled me to understand many of the methods and much of the work that otherwise would have gone on unknown to me. Early in the harvest an aspen cluster far downstream was cut. Every tree in this cluster and every near-by aspen was felled, dragged to the brook, and in this, with wrestling, pushing, and pulling, taken upstream through shallow water,—for most mountain streams are low during the autumn. In the midst of this work the entrance or inlet of the canal was blocked and the bow dam was cut. The water in the brook was almost doubled

THE DEAD-WOOD DAM, LOOKING SOUTH

The Beaver's Engineering

in volume by the closing of the canal, thereby making the transportation of aspens upstream less laborious.

When the downstream aspens at last reposed in a pile beside the house, harvesting was briskly begun in the aspens along the shore of the new pond. Then came another surprise. The bow dam was repaired, and the canal not only opened, but enlarged so that almost all the water in the brook was diverted into the canal, through which it flowed into the new pond.

The aspens cut on the shore of the new pond were floated across it, then dragged up the canal into the old pond. Evidently the beavers not only had again turned the water into the canal that they might use it in transportation, but also had increased the original volume of water simply to make this transportation of the aspens as easy as possible.

Their new works enabled the colonists to procure nearly five hundred aspens for the winter. All these were taken up the new canal, dragged over the bow and the main dams, and piled in the water by the house. In addition to these, the

In Beaver World

aspens brought from downstream made the total of the harvest seven hundred and thirty-two trees; and with these went several hundred small willows. Altogether these made a large green brush-pile that measured more than a hundred feet in circumference, and after it settled averaged four feet in depth. This was the food-supply for the oncoming winter. The upper surface of this stood about one foot above the surface of the water.

Five years after the completion of this deadwood dam it was so overgrown with willows and grass that the original material — the dead tree-trunks that formed the major portion of it — was completely covered over. The new pond was used but one season. All the aspens that were made available by the dam of the pond were cut in one harvest. The place is now abandoned, old ponds and new.

The Ruined Colony

The Ruined Colony

TWENTY-SIX years ago, while studying glaciation on the slope of Long's Peak, I came upon a cluster of eight beaver houses. These crude conical mud huts were in a forest pond far up on the mountainside. In this colony of our first engineers were so many things of interest that the fascinating study of the dead Ice King's ruins and records was indefinitely given up in order to observe Citizen Beaver's works and ways.

A pile of granite boulders on the edge of the pond stood several feet above the water-level, and from the top of these the entire colony and its operations could be seen. On these I spent days observing and enjoying the autumnal activities of Beaverdom.

It was the busiest time of the year for these industrious folk. General and extensive preparations were now being made for the long winter amid the mountain snows. A harvest of scores of trees was being gathered and work on a new house was in progress, while the old houses were receiving repairs. It was a serene autumn day when I came into the picturesque village of these primitive people. The aspens were golden, the willows rusty, the grass tanned, and the pines were purring in the easy air.

The colony-site was in a small basin amid morainal débris at an altitude of nine thousand feet above the sea-level. I at once christened it the Moraine Colony. The scene was utterly wild. Peaks of crags and snow rose steep and high above all; all around crowded a dense evergreen forest of pine and spruce. A few small swamps reposed in this forest, while here and there in it bristled several gigantic windrows of boulders. A ragged belt of aspens surrounded the several ponds and separated the pines and spruces from the fringe of water-loving willows along the shores. There were three large ponds in succession and below these a number of smaller ones. The dams that formed the large ponds were willow-grown, earthy structures about four feet in height, and all sagged downstream. The houses were grouped in the middle pond, the largest one, the dam of

The Ruined Colony

which was more than three hundred feet long. Three of these lake dwellings stood near the upper margin, close to where the brook poured in. The other five were clustered by the outlet, just below which a small willow-grown, boulder-dotted island lay between the divided waters of the stream.

A number of beavers were busy gnawing down aspens, while others cut the felled ones into sections, pushed and rolled the sections into the water, and then floated them to the harvest piles, one of which was being made beside each house. Some were quietly at work spreading a coat of mud on the outside of each house. This would freeze and defy the tooth and claw of the hungriest or the strongest predaceous enemy. Four beavers were leisurely lengthening and repairing a dam. A few worked singly, but most of them were in groups. All worked quietly and with apparent deliberation, but all were in motion, so that it was a busy scene. " To work like a beaver!" What a stirring exhibition of beaver industry and forethought I viewed from my boulder-pile!

At times upward of forty of them were in

sight. Though there was a general coöperation, yet each one appeared to do his part without orders or direction. Time and again a group of workers completed a task, and without pause silently moved off, and began another. Everything appeared to go on mechanically. It produced a strange feeling to see so many workers doing so many kinds of work effectively and automatically. Again and again I listened for the superintendent's voice; constantly I watched to see the overseer move among them; but I listened and watched in vain. Yet I feel that some of the patriarchal fellows must have carried a general plan of the work, and that during its progress orders and directions that I could not comprehend were given from time to time.

The work was at its height a little before midday. Nowadays it is rare for a beaver to work in daylight. Men and guns have prevented daylight workers from leaving descendants. These not only worked but played by day. One morning for more than an hour there was a general frolic, in which the entire population appeared to take part. They raced, dived, crowded in general

The Ruined Colony

mix-ups, whacked the water with their tails, wrestled, and dived again. There were two or three play-centres, but the play went on without intermission, and as their position constantly changed, the merrymakers splashed water all over the main pond before they calmed down and in silence returned to work. I gave most attention to the harvesters, who felled the aspens and moved them, bodily or in sections, by land and water to the harvest piles. One tree on the shore of the pond, which was felled into the water, was eight inches in diameter and fifteen feet high. Without having even a limb cut off, it was floated to the nearest harvest pile. Another, about the same size, which was procured some fifty feet from the water, was cut into four sections and its branches removed; then a single beaver would take a branch in his teeth, drag it to the water, and swim with it to a harvest pile. But four beavers united to transport the largest section to the water. They pushed with fore paws, with breasts, and with hips. Plainly it was too heavy for them. They paused. "Now they will go for help," I said to myself, "and I shall find out who

In Beaver World

the boss is." But to my astonishment one of them began to gnaw the piece in two, and two more began to clear a narrow way to the water, while the fourth set himself to cutting down another aspen. Good roads and open waterways are the rule, and perhaps the necessary rule, of beaver colonies.

I became deeply interested in this colony, which was situated within two miles of my cabin, and its nearness enabled me to be a frequent visitor and to follow closely its fortunes and misfortunes. About the hut-filled pond I lingered when it was covered with winter's white, when fringed with the gentian's blue, and while decked with the pond-lily's yellow glory.

Fire ruined it during an autumn of drouth. One morning, while watching from the boulder-pile, I noticed an occasional flake of ash dropping into the pond. Soon smoke scented the air, then came the awful and subdued roar of a forest fire. I fled, and from above the timber-line watched the storm-cloud of black smoke sweep furiously forward, bursting and closing to the terrible leaps of red and tattered flames. Before noon several thou-

The Ruined Colony

sand acres of forest were dead, all leaves and twigs were in ashes, all tree-trunks blistered and blackened.

The Moraine Colony was closely embowered in a pitchy forest. For a time the houses in the water must have been wrapped in flames of smelter heat. Could these mud houses stand this? The beavers themselves I knew would escape by sinking under the water. Next morning I went through the hot, smoky area and found every house cracked and crumbling; not one was inhabitable. Most serious of all was the total loss of the uncut food-supply, when harvesting for winter had only begun.

Would these energetic people starve at home or would they try to find refuge in some other colony? Would they endeavor to find a grove that the fire had missed and there start anew? The intense heat had consumed almost every fibrous thing above the surface. The piles of garnered green aspen were charred to the water-line; all that remained of willow thickets and aspen groves were thousands of blackened pickets and points, acres of coarse charcoal stubble. It

In Beaver World

was a dreary, starving outlook for my furred friends.

I left the scene to explore the entire burned area. After wandering for hours amid ashes and charcoal, seeing here and there the seared carcass of a deer or some other wild animal, I came upon a beaver colony that had escaped the fire. It was in the midst of several acres of swampy ground that was covered with fire-resisting willows and aspens. The surrounding pine forest was not dense, and the heat it produced in burning did no damage to the scattered beaver houses.

From the top of a granite crag I surveyed the green scene of life and the surrounding sweep of desolation. Here and there a sodden log smouldered in the ashen distance and supported a tower of smoke in the still air. A few miles to the east, among the scattered trees of a rocky summit, the fire was burning itself out; to the west the sun was sinking behind crags and snow; near by, on a blackened limb, a south-bound robin chattered volubly but hopelessly.

While I was listening, thinking, and watching, a mountain lion appeared and leaped lightly upon

The Ruined Colony

a block of granite. He was on my right, about one hundred feet away and about an equal distance from the shore of the nearest pond. He was interested in the approach of something. With a nervous switching of his tail he peered eagerly forward over the crown of the ridge just before him, and then crouched tensely and expectantly upon his rock.

A pine tree that had escaped the fire screened the place toward which the lion looked and where something evidently was approaching. While I was trying to discover what it could be, a coyote trotted into view. Without catching sight of the near-by lion, he suddenly stopped and fixed his gaze upon the point that so interested the crouching beast. The mystery was solved when thirty or forty beavers came hurrying into view. They had come from the ruined Moraine Colony.

I thought to myself that the coyote, stuffed as he must be with the seared flesh of fire-roasted victims, would not attack them; but a lion wants a fresh kill for every meal, and so I watched the movements of the latter. He adjusted his feet a trifle and made ready to spring. The beavers

In Beaver World

were getting close; but just as I was about to shout to frighten him, the coyote leaped among them and began killing.

In the excitement of getting off the crag I narrowly escaped breaking my neck. Once on the ground, I ran for the coyote, shouting wildly to frighten him off; but he was so intent upon killing that a violent kick in the ribs first made him aware of my presence. In anger and excitement he leaped at me with ugly teeth as he fled. The lion had disappeared, and by this time the beavers in the front ranks were jumping into the pond, while the others were awkwardly speeding down the slope. The coyote had killed three. If beavers have a language, surely that night the refugees related to their hospitable neighbors some thrilling experiences.

The next morning I returned to the Moraine Colony over the route followed by the refugees. Leaving their fire-ruined homes, they had followed the stream that issued from their ponds. In places the channel was so clogged with fire wreckage that they had followed alongside the water rather than in it, as is their wont. At one

The Ruined Colony

place they had hurriedly taken refuge in the stream. Coyote tracks in the scattered ashes explained this. But after going a short distance they had climbed from the water and again traveled the ashy earth.

Beavers commonly follow water routes, but in times of emergency or in moments of audacity they will journey overland. To have followed this stream down to its first tributary, then up this to where the colony in which they found refuge was situated, would have required four miles of travel. Overland it was less than a mile. After following the stream for some distance, at just the right place they turned off, left the stream, and dared the overland dangers. How did they know the situation of the colony in the willows, or that it had escaped fire, and how could they have known the shortest, best way to it?

The morning after the arrival of the refugees, work was begun on two new houses and a dam, which was about sixty feet in length and built across a grassy open. Green cuttings of willow, aspen, and alder were used in its construction.

In Beaver World

Not a single stone or handful of mud was used. When completed it appeared like a windrow of freshly raked shrubs. It was almost straight, but sagged a trifle downstream. Though the water filtered freely through, it flooded the flat above. As the two new houses could not shelter all the refugees, it is probable that some of them were sheltered in bank tunnels, while room for others may have been found in the old houses.

That winter the colony was raided by some trappers; more than one hundred pelts were secured, and the colony was left in ruins and almost depopulated.

The Moraine Colony site was deserted for a long time. Eight years after the fire I returned to examine it. The willow growth about the ruins was almost as thrifty as when the fire came. A growth of aspen taller than one's head clung to the old shore-lines, while a close seedling growth of lodge-pole pine throve in the ashes of the old forest. One low mound, merry with blooming columbine, was the only house ruin to be seen.

The ponds were empty and every dam was broken. The stream, in rushing unobstructed

The Ruined Colony

through the ruins, had eroded deeply. This erosion revealed the records of ages, and showed that the old main dam had been built on the top of an older dam and a sediment-filled pond. The second dam was on top of an older one still. In the sediment of the oldest — the bottom pond — I found a spearhead, two charred logs, and the skull of a buffalo. Colonies of beaver, as well as those of men, are often found upon sites that have a tragic history. Beavers, with Omar, might say, —

"When you and I behind the veil are past,
Oh but the long long while the world shall last."

The next summer, 1893, the Moraine site was resettled. During the first season the colonists spent their time repairing dams and were content to live in holes. In autumn they gathered no harvest, and no trace of them could be found after the snow; so it is likely that they had returned to winter in the colony whence they had come. But early in the next spring there were reinforced numbers of them at work establishing a permanent settlement. Three dams were repaired, and in the autumn many of the golden

In Beaver World

leaves that fell found lodgment in the fresh plaster of two new houses.

In the new Moraine Colony one of the houses was torn to pieces by some animal, probably a bear. This was before Thanksgiving. About midwinter a prospector left his tunnel a few miles away, came to the colony and dynamited a house, and "got seven of them." Next year two houses were built on the ruins of the two just fallen. That year's harvest-home was broken by deadly attacks of enemies. In gathering the harvest the beavers showed a preference for some aspens that were growing in a moist place about one hundred feet from the water. Whether it was the size of these or their peculiar flavor that determined their election in preference to nearer ones, I could not determine. One day, while several beavers were cutting here, they were surprised by a mountain lion which leaped upon and killed one of the harvesters. The next day the lion surprised and killed another. Two or three days later a coyote killed one on the same blood-stained spot, and then overtook and killed two others as they fled for the water. I could not see these deadly attacks

The Ruined Colony

from the boulder-pile, but in each case the sight of flying beavers sent me rushing upon the scene, where I beheld the cause of their desperate retreat. But despite dangers they persisted until the last of these aspens was harvested. During the winter the bark was eaten from these, and the next season their clean wood was used in the walls of a new house.

One autumn I had the pleasure of seeing some immigrants pass me *en route* for a new home in the Moraine Colony. Of course they may have been only visitors, or have come temporarily to assist in the harvesting; but I like to think of them as immigrants, and a number of things testified that immigrants they were. One evening I had been lying on a boulder by the stream below the colony, waiting for a gift from the gods. It came. Out of the water within ten feet of me scrambled the most patriarchal, as well as the largest, beaver that I have ever seen. I wanted to take off my hat to him, I wanted to ask him to tell me the story of his life, but from long habit I simply lay still and watched and thought in silence. He was making a portage round a cascade. As he scrambled up over the rocks, I noticed that he had but two fingers on his right hand. He was followed, in single file, by four others; one of these was minus a finger on the left hand. The next morning I read that five immigrants had arrived in the Moraine Colony. They had registered their footprints in the muddy margin of the lower pond. Had an agent been sent to invite these colonists, or had they come out of their own adventurous spirit? The day following their arrival I trailed them backward in the hope of learning whence they came and why they had moved. They had traveled in the water most of the time; but in places they had come out on the bank to go round a waterfall or to avoid an obstruction. Here and there I saw their tracks in the mud and traced them to a beaver settlement in which the houses and dams had been recently wrecked. A near-by rancher told me that he had been "making it hot" for all beavers in his meadow. During the next two years I occasionally saw this patriarchal beaver or his tracks thereabout.

It is the custom among old male beavers to idle away two or three months of each summer

The Ruined Colony

in exploring the neighboring brooks and streams, but they never fail to return in time for autumn activities. It thus becomes plain how, when an old colony needs to move, some one in it knows where to go and the route to follow.

The Moraine colonists gathered an unusually large harvest during the autumn of 1909. Seven hundred and thirty-two sapling aspens and several hundred willows were massed in the main pond by the largest house. This pile, which was mostly below the water-line, was three feet deep and one hundred and twenty-four feet in circumference. Would a new house be built this fall? This unusually large harvest plainly told that either children or immigrants had increased the population of the colony. Of course, a hard winter may also have been expected.

No; they were not to build a new house, but the old house by the harvest pile was to be enlarged. One day, just as the evening shadow of Long's Peak had covered the pond, I peeped over a log on top of the dam to watch the work. The house was only forty feet distant. Not a ripple stirred among the inverted peaks and pines in the clear, shadow-enameled pond. A lone beaver rose quietly in the scene from the water near the house. Swimming noiselessly, he made a circuit of the pond. Then for a time, and without any apparent purpose, he swam back and forth over a short, straight course; he moved leisurely, and occasionally made a shallow, quiet dive. He did not appear to be watching anything in particular or to have anything special on his mind. Yet his eyes may have been scouting for enemies and his mind may have been full of house plans. Finally he dived deeply, and the next I saw of him he was climbing up the side of the house addition with a pawful of mud.

By this time a number of beavers were swimming in the pond after the manner of the first one. Presently all began to work. The addition already stood more than two feet above the waterline. The top of this was crescent-shaped and was about seven feet long and half as wide. It was made mostly of mud, which was plentifully reinforced with willow cuttings and aspen sticks. For a time all the workers busied themselves in carrying mud and roots from the bottom of the

The Ruined Colony

pond and placing these on the slowly rising addition. Eleven were working at one time. By and by three swam ashore, each in a different direction and each a few seconds apart. After a minute or two they returned from the shore, each carrying or trailing a long willow. These were dragged to the top of the addition, laid down, and trampled in the mud. Meantime the mud-carriers kept steadily at their work; again willows were brought, but this time four beavers went, and, as before, each was independent of the others. I did not see how this work could go on without some one bossing the thing, but I failed to detect any beaver acting as overseer. While there was general coöperation, each acted independently most of the time and sometimes was apparently oblivious of the others. These beavers simply worked, slowly, silently, and steadily; and they were still working away methodically and with dignified deliberation when darkness hid them.

Beaver Pioneers

Beaver Pioneers

I OFTEN wish that an old beaver neighbor of mine would write the story of his life. Most of the time for eighteen years his mud hut was among the lilies of Lily Lake, Estes Park, Colorado. He lived through many wilderness dangers, escaped the strategy of trappers, and survived the dangerous changes that come in with the home-builder. His life was long, stirring, and adventurous. If, in the first chapter of his life-story, he could record some of the strong, thrilling experiences which his ancestors must have related to him, his book would be all the better.

"Flat-top," my beaver neighbor, was a pioneer and a colony-founder. It is probable that he was born in a beaver house on Wind River, and it is likely that he spent the first six years of his life along this crag and aspen bordered mountain stream. The first time I saw him he was leading an emigrant party out of this stream's steep-

In Beaver World

walled upper course. He and his party settled, or rather resettled, Lily Lake.

Flat-top was the name I gave him because of his straight back. In most beaver the shoulders swell plumply above the back line after the outline of the grizzly bear. Along with this peculiarity, which enabled me to be certain of his presence, was another. This was his habit of gnawing trees off close to the earth when he felled them. The finding of an occasional low-cut stump assured me of his presence during the periods I failed to see him.

The first beaver settlement in the lake appears to have been made in the early seventies, long before Flat-top was born, by a pair of beaver who were full of the pioneer spirit. These settlers apparently were the sole survivors of a large party of emigrants who tried to climb the rugged mountains to the lake, having been driven from their homes by encroaching human settlers. After a long, tedious journey, full of hardships and dangers, they climbed into the lake that was to them, for years, a real promised land.

Driven from Willow Creek, they set off up-

Beaver Pioneers

stream in search of a new home, probably without knowing of Lily Lake, which was five miles distant and two thousand feet up a steep, rocky mountain. These pilgrims had traveled only a little way upstream when they found themselves the greater portion of the time out of water. This was only a brook at its best and in most places it was such a shallow, tiny streamlet that in it they could not dive beyond the reach of enemies or even completely cool themselves. In stretches the water spread thinly over a grassy flat or a smooth granite slope; again it was lost in the gravel; or, murmuring faintly, pursued its way out of sight beneath piles of boulder, — marbles shaped by the Ice King. Much of the time they were compelled to travel upon land exposed to their enemies. Water-holes in which they could escape and rest were long distances apart.

This plodding, perilous five-mile journey which the beaver made up the mountain to the lake would be easy and care-free for an animal with the physical make-up of a bear or a wolf, but with the beaver it is not surprising that only two of the emigrants survived this supreme trial and

In Beaver World

escaped the numerous dangers of the pilgrimage.

Lily Lake is a shallow, rounded lily garden that reposes in a glacier meadow at an altitude of nine thousand feet; its golden pond-lilies often dance among reflected snowy peaks, while over it the granite crags of Lily Mountain rise several hundred feet. A few low, sedgy, grassy acres border half the shore, while along the remainder are crags, aspen groves, willow-clumps, and scattered pines. Its waters come from springs in its western margin and overflow across a low grassy bar on its curving eastern shore.

It was autumn when these beaver pioneers came to Lily Lake's primitive and poetic border. The large green leaves of the pond-lily rested upon the water, while from the long green stems had fallen the sculptured petals of gold; the willows were wearing leaves of brown and bronze, and the yellow tremulous robes of the aspens glowed in the golden sunlight.

These fur-clad pioneers made a dugout — a hole in the bank — and busily gathered winter food until stopped by frost and snow; then, almost

Beaver Pioneers

care-free, they dozed away the windy winter days while the lake was held in waveless ice beneath the drifting snow.

The next summer a house was built in the lily pads near the shore. Here a number of children were born during the few tranquil years that followed. These times came to an end one bright midsummer day. Lord Dunraven had a ditch cut in the outlet rim of the lake with the intention of draining it that his fish ponds, several miles below in his Estes Park game-preserve, might have water. A drouth had prevailed for several months, and a new water-supply must be had or the fish ponds would go dry. The water poured forth through the ditch, and the days of the colony appeared to be numbered.

A beaver must have water for safety and for the ease of movement of himself and his supplies. He is skillful in maintaining a dam and in regulating the water-supply; these two things require much of his time. In Lily Lake the dam and the water question had been so nicely controlled by nature that with these the colonists had had nothing to do. However, they still knew

In Beaver World

how to build dams, and water-control had not become a lost art. The morning after the completion of the drainage ditch, a man was sent up to the lake to find out why the water was not coming down. A short time after the ditch-diggers had departed, the lowering water had aroused the beaver, who had promptly placed a dam in the mouth of the ditch. The man removed this dam and went down to report. The beaver speedily replaced it. Thrice did the man return and destroy their dam, but thrice did the beaver promptly restore it.

The dam-material used in obstructing the ditch consisted chiefly of the peeled sticks from which the beaver had eaten the bark in winter; along with these were mud and grass. The fourth time that the ditch guard returned, he threw away all the material in the dam and then set some steel traps in the water by the mouth of the ditch. The first two beaver who came to reblockade the ditch were caught in these traps and drowned while struggling to free themselves. Other beaver heroically continued the work that these had begun. The cutting down of saplings

HOUSE IN LILY LAKE

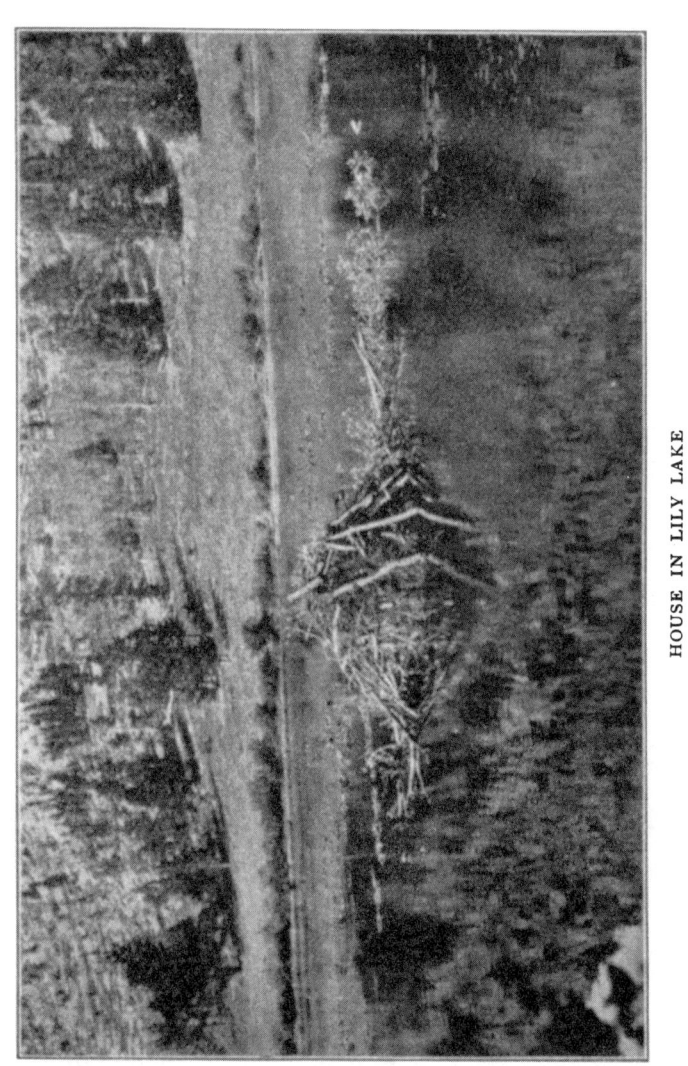

Beaver Pioneers

and the procuring of new material made their work slow, very slow, in the face of the swiftly escaping water; when the ditch was at last obstructed, a part of the material which formed this new dam consisted of the traps and the dead bodies of the two beaver who had bravely perished while trying to save the colony.

The ditch guard returned with a rifle, and came to stay. The first beaver to come within range was shot. The guard again removed the dam, made a fire about twenty feet from the ditch, and planned to spend the night on guard, rifle in hand. Toward morning he became drowsy, sat down by the fire, heard the air in the pines at his back, watched the star-sown water, and finally fell asleep. While he thus slept, with his rifle across his lap, the beaver placed another — their last — obstruction before the outrushing water.

On awakening, the sleeper tore out the dam and stood guard over the ditch. All that afternoon a number of beaver hovered about, watching for an opportunity to stop the water again. Their opportunity never came, and three who ventured too near the rifleman gave up their lives, — reddening the clear water with their life-blood in vain.

The lake was drained, and the colonists abandoned their homes. One night, a few days after the final attempt to blockade the ditch, an unwilling beaver emigrant party climbed silently out of the uncovered entrance of their house and made their way quietly, slowly, beneath the stars, across the mountain, descending thence to Wind River, where they founded a new colony.

Winter came to the old lake-bed, and the lily roots froze and died. The beaver houses rapidly crumbled, and for a few years the picturesque ruins of the beaver settlement, like many a settlement abandoned by man, stood pathetically in the midst of wilderness desolation. Slowly the water rose to its old level in the lake, as the outlet ditch gradually filled with swelling turf and drifting sticks and trash. Then the lilies came back with rafts of green and boats of gold to enliven this lakelet of repose.

One autumn morning, while returning to my cabin after a night near the stars on Lily Moun-

Beaver Pioneers

tain, I paused on a crag to watch the changing morning light down Wind River Cañon. While thus engaged, Flat-top and a party of colonists came along a game trail within a few yards of me, evidently bound for the lake, which was only a short distance away. I silently followed them. This was my introduction to Flat-top.

On the shore these seven adventurers paused for a moment to behold the scene, or, possibly, to dream of empire; then they waddled out into the water and made a circuit of the lake. Probably Flat-top had been here before as an explorer. Within two hours after their arrival these colonists began building for a permanent settlement.

It was late to begin winter preparation. The clean, white aspens had shed their golden leaves and stood waiting to welcome the snows. This lateness may account for the makeshift of a hut which the colonists constructed. This was built against the bank with only one edge in the water; the entrance to it was a twelve-foot tunnel that ended in the lake-bottom where the water was two feet deep.

The beaver were collecting green aspen and

In Beaver World

willow cuttings in the water by the tunnel-entrance when the lake froze over. Fortunately for the colonists, with their scanty supply of food, the winter was a short one, and by the first of April they were able to dig the roots of water plants along the shallow shore where the ice had melted. One settler succumbed during the winter, but by summer the others had commenced work on a permanent house, which was completed before harvest time.

I had a few glimpses of the harvest-gathering and occasionally saw Flat-top. One evening, while watching the harvesters, I saw three new workers. Three emigrants — from somewhere — had joined the colonists. A total of fifteen, five of whom were youngsters, went into winter quarters, — a large, comfortable house, a goodly supply of food, and a location off the track of trappers. The cold, white days promised only peace. But an unpreventable catastrophe came before the winter was half over.

One night a high wind began to bombard the ice-bound lake with heavy blasts. The force of these intermittent gales suggested that the wind

Beaver Pioneers

was trying to dislodge the entire ice covering of the lake; and indeed that very nearly happened.

Before the crisis came, I went to the lake, believing it to be the best place to witness the full effects of this most enthusiastic wind. Across the ice the gale boomed, roaring in the restraining forest beyond. These broken rushes set the ice vibrating and the water rolling and swelling beneath. During one of these blasts the swelling water burst the ice explosively upward in a fractured ridge entirely across the lake. In the next few minutes the entire surface broke up, and the wind began to drive the cakes upon the windward shore.

A large flatboat cake was swept against the beaver house, sheared it off on the water-line, and overturned the conelike top into the lake. The beaver took refuge in the tunnel which ran beneath the lake-bottom. This proved a death-trap, for its shore end above the water-line was clogged with ice. As the lake had swelled and surged beneath the beating of the wind, the water had gushed out and streamed back into the tunnel again and again, until ice formed in and closed

In Beaver World

the outer entrance. Against this ice four beaver were smothered or drowned. I surmised the tragedy but was helpless to prevent it. Meanwhile the others doubled back and took refuge upon the ruined stump of their home. From a clump of near-by pines I watched this wild drama.

Less than half an hour after the house was wrecked, these indomitable animals began to rebuild it. Lashed by icy waves, beaten by the wind, half-coated with ice, these home-loving people strove to rebuild their home. Mud was brought from the bottom of the pond and piled upon the shattered foundation. This mud set — froze — almost instantly on being placed. They worked desperately, and from time to time I caught sight of Flat-top. Toward evening it appeared possible that the house might be restored, but, just as darkness was falling, a roaring gust struck the lake and a great swell threw the new part into the water.

The colonists gave up the hopeless task and that night fled down the mountain. Two were killed before they had gone a quarter of a mile. Along the trail were three other red smears upon

Beaver Pioneers

the crusted snow; each told of a death and a feast upon the wintry mountain-side among the solemn pines. Flat-top with five others finally gained the Wind River Colony, from which he had led his emigrants two years before.

One day the following June, while examining the lilies in the lake, I came upon a low, freshly cut stump;—Flat-top had returned. A number of colonists were with him and all had come to stay.

All sizable aspen that were within a few yards of the water had been cut away, but at the southwest corner of the lake, about sixty feet from the shore, was an aspen thicket. Flat-top and his fellow workers cut a canal from the lake through a low, sedgy flat into this aspen thicket. The canal was straight, about fourteen inches deep and twenty-six inches wide. Its walls were smoothly cut and most of the excavated material was piled evenly on one side of the canal and about eight inches from it. It had an angular, mechanical appearance, and suggested the work not of a beaver, but of man, and that of a very careful man too.

Down this canal the colonists floated the timbers used in building their two houses. On the completion of the houses, the home-builders returned to the grove and procured winter supplies. In most cases the small aspen were floated to the pile between the houses with an adept skill, without severing the trunk or cutting off a single limb.

The colonists had a few years of ideal beaver life. One summer I came upon Flat-top and a few other beaver by the brook that drains the lake, and at a point about half a mile below its outlet. It was along this brook that Flat-top's intrepid ancestors had painfully climbed to establish the first settlement in the lake. Commonly each summer several beaver descended the mountain and spent a few weeks of vacation along Wind River. Invariably they returned before the end of August; and autumn harvest-gathering usually began shortly after their return.

Year after year the regularly equipped trappers passed the lake without stopping. The houses did not show distinctly from the trail, and the trappers did not know that there were beaver in this place. But this peaceful, populous lake was

Beaver Pioneers

not forever to remain immune from the wiles of man, and one day it was planted with that barbaric, cruel torture-machine, the steel trap.

A cultured consumptive, who had returned temporarily to nature, was boarding at a ranch house several miles away. While out riding he discovered the colony and at once resolved to depopulate it. The beaver ignored his array of traps until he enlisted the services of an old trapper, whose skill sent most of the beaver to their death before the sepia-colored catkins appeared upon the aspens. Flat-top escaped.

The ruinous raid of the trappers was followed by a dry season, and during the drouth a rancher down the mountain came up prospecting for water. He cut a ditch in the outlet ridge of the lake, and out gushed the water. He started home in a cheerful mood, but long before he arrived, the "first engineers" had blocked his ditch. During the next few days and nights the rancher made many trips from his house to the lake, and when he was not in the ditch, swearing, and opening it, the beaver were in it shutting off the water.

In Beaver World

From time to time I dropped around to see the struggle, one day coming upon the scene while the beaver were completing a blockade. For a time the beaver hesitated; then they partly resumed operations and carried material to the spot, but without showing themselves entirely above water. When it appeared that they must have enough to complete the blockade, I advanced a trifle nearer so as to have a good view while they placed the accumulated material. For a time not a beaver showed himself. By and by an aged one climbed out of the water, pretending not to notice me, and deliberately piled things right and left until he had completed the ditch-damming to his satisfaction. This act was audacious and truly heroic. The hero was Flat-top.

In this contest with the rancher, the beaver persisted and worked so effectively that they at last won and saved their homes, in the face of what appeared to be an unconquerable opposition.

A little while after this incident, a home-seeker came along, and, liking the place, built a cabin in a clump of pines close to the southern shore. Though he was a gray old man without a family,

Beaver Pioneers

I imagined he would exterminate the beaver and looked upon him with a lack of neighborly feeling.

Several months went by, and I had failed to call upon him, but one day while passing I heard him order a trapper off the place. This order was accompanied by so strong a declaration of principles — together with a humane plea for the life of every wild animal — that I made haste to call that evening.

One afternoon in a pine thicket, close to the lake-shore, I came upon two gray wolves, both devouring beaver, which had met their death while harvesting aspens for winter. The following spring I had a more delightful glimpse of life in the wilds. Within fifty feet of the lake-shore stood a large pine stump that rose about ten feet from the ground. Feeling that I should escape notice if I sat still on the top, I climbed up. Though it was mid-forenoon, the beaver came out of the lake and wandered about nibbling here and there at the few green plants of early spring. They did not detect me. They actually appeared to enjoy themselves. This is the

In Beaver World

only time that I ever saw a beaver fully at ease and apparently happy on land. In the midst of their pleasures, a flock of mountain sheep came along and mingled with them. The beaver paused and stared; now and then a sheep would momentarily stare at a beaver, or sniff the air as though he did not quite like beaver odor. In less than a minute the flock moved on, but just as they started, a beaver passed in front of the lead ram, who made a playful pretense of a butt at him; to this the beaver paid not the slightest heed.

During the homesteader's second summer he concluded to raise the outlet ridge, deepen the water, and make a fish pond of the lake. Being poor, he worked alone with wheelbarrow and shovel. The beaver evidently watched the progress of the work, and each morning their fresh footprints showed in the newly piled earth. Shortly before the dam was completed, the homesteader was called away for a few days, and on his return he was astonished to find that the beaver had completed his dam! The part made by the beaver suited him as to height and length,

Beaver Pioneers

so he covered it over with earth and allowed it to remain. His work in turn was inspected and apparently approved by the beaver.

How long does a beaver live? Trappers say from fifteen to fifty years. I had glimpses of Flat-top through eighteen years, and he must have been not less than four years of age when I first met him. This would make his age twenty-two years; but he may have been six years of age — he looked it — the morning he first led emigrants into Lily Lake; and he may have lived a few years after I saw him last. But only the chosen few among the beaver can succeed in living as long as Flat-top. The last time I saw him was the day he dared me and blockaded the drain ditch and stopped the outrushing water.

Flat-top has vanished, and the kind old homesteader has gone to his last long sleep; but the lake still remains, and still there stands a beaver house among the pond-lilies.

The Colony in Winter

The Colony in Winter

IN the Medicine Bow Mountains one December day, I came upon a beaver house that was surrounded by a pack of wolves. These beasts were trying to break into the house. Apparently an early autumn snow had blanketed the house and thus prevented its walls from freezing. The soft condition of the walls, along with the extreme hunger of the wolves, led to this assault. Two of these animals were near the top of the house clawing away at a rapid rate. Now and then one of the sticks or poles in the house-wall was encountered, and at this the wolf would bite and tear furiously. Occasionally one of the wolves caught a resisting stick in his teeth, and, leaning back, shook his head, endeavoring with all his might to tear it out. A number of wolves lay about expectant; a few sat up eagerly on haunches, while others moved about snarling, driving the others off a few yards, to be in turn driven off themselves. Shortly before they discovered me, there was a fierce fight on top of the house, in which several mixed.

Even though they had broken into the house, it would have availed them nothing, for in this, as in all old colonies, there were safety tunnels from the house which extended beneath the pond to points on shore. In these tunnels the beaver find safety, if by any means the house is ruined. Although carnivorous animals are fond of beaver flesh, they rarely take the useless trouble of digging into a house. Occasionally a wolverine or a bear may dig into a thin-walled house or one not frozen, then, after breaking in, lie in wait, and endeavor to make a capture while the beaver are repairing the hole. Beaver are more secure from enemies during the winter than at any other time. It is while felling a tree far from the water or while following a shallow stream that most beaver are captured by their enemies.

Many a time in winter I have made a pleasant visit to a beaver colony. One day, a few hours after a heavy snowfall, I came out of a dark forest and stood for a time on the edge of the snow-covered pond. Around were the firs and spruces

The Colony in Winter

of the forest, moveless as statues and each a pointed cone of snow. Around the small snowy plain of the pond, the drooping snow-entangled willows held their heads together in contented and thoughtful silence. Everything was serene.

A clean fox track led from the woods in a straight line across the snowy surface of the pond to the house, which stood near the centre of this smooth white opening. The tracks encircled the house and ascended to the top of it, where the record imprinted in the snow told that here he watchfully rested. Descending, he had sniffed at the bushy tips of the winter food-pile that thrust up through the ice, then crossed the dam to plunge into the snowy tangle of willows.

Water was still pouring and gurgling down a steep beaver slide. This was ice-and-snow-covered except at two points where the swift splashing water dashed intermittently from a deep icy vent. While I was examining the beauty of the up-building icy buttresses by one of the vents, a water-ouzel came forth and alighted almost within reach. I stood still. After giving a few of his nodding bows, he reëntered the vent. Presently he emerged

In Beaver World

from the lower vent and, alighting upon an ice-coated boulder, indifferent to the gray sky from which scattered flakes were slowly falling and despite a temperature of five below zero, he sang low and sweetly for several seconds.

Beaver do not surrender themselves to the confines of a house and pond until cold solidly covers the pond with a roof of ice. The time of this is commonly about the first of December, but the date is of course, in a measure, dependent upon latitude, altitude, and the peculiar weather conditions of each year. Most beaver return to the old colony, or start a new one by the first of September. They have had a merry rambling summer and energetically take hold to have the house and dam ready and a harvest stored by the time winter begins.

But they are not always ready. Enemies may harass them, low water delay them, or an unusually early winter or even a heavy snow may so hamper them that, despite greatest effort, the ice puts a time lock upon the pond and closes them in for the winter without sufficient supplies.

Early one October an early snowfall worked

The Colony in Winter

hardship in several colonies near my home. Fortunately the ponds were not deeply frozen, and those colonies which had aspen groves close to the water succeeded in felling and dragging in sufficient food-supplies for the winter. As snow drifted into the groves, many of the trees harvested were cut from the tops of snowdrifts, and thus left high stumps. The following summer a number of these stood four feet above the earth and presented a striking appearance alongside the sixteen-inch stumps of normal height.

One of these storm-caught colonies fared badly. The inhabitants were obliged to go a long distance from the water for trees, and their all too scanty harvest was gathered with some loss of life. Apparently both wolves and lions discovered the unfortunate predicament of the harvesters, and lay in wait to catch them as they floundered slowly through the snow. The following winter these colonists tunneled through the bottom — perhaps the least frozen part of the dam — and came forth for food long before the break-up of the ice. The water drained from the pond, and after the ice had melted, the bottom of the pond revealed a torn-up condition as though the starving winter inmates had dug out for food every root and rootstock to be found in the bottom.

While visiting ponds at the beginning of winter, I have many times noticed that, shortly after the pond was solidly frozen over, a hole was made through the dam just below the water-surface of the pond. This lowered the water-level two inches or more. Did this slight lowering of the water have to do with the ventilation of the ice-covered pond, or was it to put a check on deep freezing, or for both purposes?

In the majority of cases these holes were made from ponds which, during the winter, received but a meagre inflow of fresh water. Naturally, ponds receiving a strong inflow of water would be better ventilated, and would freeze less swiftly and deeply than those whose waters became stagnant. This drawing-off of water after a few inches of ice had formed, would, in some places, despite the settling of the ice, form an air blanket that would delay freezing, and thus possibly prevent the ice from forming so thickly. The air admitted by drawing off the water would be in-

The Colony in Winter

closed beneath the ice, and might thus be helpful to the beaver inclosed in house and pond. In only a few cases were these holes made from ponds which had subway tunnels,— tunnels which run from alongside the house through the bottom of the pond to a point above water-level on the shore. In a few instances the beaver, I do not know how many, came out of this hole, cut and ate a few twigs, and then returned and closed it. Twice this was used as a way out by beaver who emerged and went to other colonies. In one case the beaver entered the other pond by making a hole through the dam. In the other they entered the pond through a subway tunnel. While these holes which lower the pond-level may have chiefly to do with ventilation, or may be for the purpose of putting a check on freezing, my evidence is not ample enough for final conclusions.

A sentence of close confinement for about a third of the year for an animal that breathes air and uses pure water, is simply one of the strange ways that work out with nature. While winter lasts, a beaver must spend his time either in the dark, ill-ventilated house or in the water of the

In Beaver World

pond. Apparently he does much sleeping and possibly has a dull time of it. No news, no visitors, and apparently nothing to do! Still a beaver has food, and when dangers surround the wild folk outside the pond's roof of glass, he would be considered a good risk for life insurance.

Although the pond is commonly covered with snow, or the ice curtained with air bubbles, there have been numerous times during which I have had clear views into the water, and could see and enjoy all that was going on within, as completely as though looking at fish or turtles through the glass walls of an aquarium. Often I have peered through the ice which covered the most used place of a winter beaver pond,— the area between the house-entrance and the food-pile. The thinness of the ice over this place was maintained by spring-water which came up through the bottom, and the beaver had so arranged their affairs that they made the best use of this shallow-freezing water. Of course most ponds are without springs.

Many a time I have seen a beaver come out of the doorway of his house and go swimming to-

The Colony in Winter

ward the food-pile with his hands against his breast. At the pile, if there was nothing small or short enough, he set to work and gnawed it off. The piece secured was taken into the doorway either in his hands or in his teeth. Afterward a beaver — the same one, I suppose — came out of the doorway, and cast the clean bone of the stick, from which the bark had been eaten, into the bottom of the pond.

When there is nothing else to do, the beaver apparently comes into the pond a few times each day for a swim. In the midst of swimming he rises at times to the under surface of the ice and, with his nose against it, exhales a quantity of air. After remaining with nose at this point a few seconds, the action of the air bubbles indicates that he is inhaling the purified air.

The rootstocks of the water-lily are sometimes dug from the bottom of the pond. At other times the beaver eats the stalks of plants that grow in the water, or digs out willow or other roots around the edge of the pond. Numbers of trout frequently lie in the water close to the doorway of a beaver house or around the food-pile. Possibly the beaver

In Beaver World

dispense tidbits of food that are liked by the trout. Occasionally grubs fall from the holes in wood from which beaver have eaten the bark. While beaver are digging in the bottom of the pond they doubtless unearth food-scraps that are welcome to trout, for these often hover in numbers on the outskirts of the muddy water which beaver roil while digging.

Although it appears that beaver have dull winters with but little to do but eat, sleep, and swim, it is probable that some of their time is spent at work. A part of their tunneling and pond-bottom canal-digging is done in winter. I have known of their extending canals in the bottom of the pond and making submarine tunnels while the pond was ice-covered.

There are times when the dam has sprung a leak and must be repaired on the inside beneath the ice. Early thaws and spring freshets sometimes wreck a dam beyond repair, or do extensive damage to the house or dam at the time when beaver enemies are likely to be at their leanest. The house and dam are sometimes ruined when the streams are so low and icy that it is not safe

The Colony in Winter

for beaver to go about. I know of two colonies that were crushed out of existence by snow-slides.

The dam is on rare occasions broken by late spring ice-jams. Sometimes the ice-cakes pile up on the dam and raise the water in the pond to such a height that it rises in the house and drives the beaver forth. A few beaver houses that are situated in places where the ice or spring floods may raise the water much above normal level are shaped to meet this trouble. The house is built higher and the room internally is twice the usual height. Thus there is space for the beaver to build a "platform bed" on the floor and thus raise themselves a foot or more above the common level. Despite all pains, floods sometimes drive beaver to the housetops.

By laying up supplies, and by the help of artificial pond, canal, and house, the beaver is able to spend his winter without hunger and with comfort and far greater safety than his neighbors. The winds may blow and blinding snow or flying limbs may endanger those outside; snow may bury the forage of bird and deer, and make the movement of beasts of prey slow and difficult; the

In Beaver World

cold may freeze and freeze and strew the wilds with lean and frozen forms; but the beaver beneath ice and snow shelter serenely spends the days with comfort and safety.

The winter, with its days long or short, never comes to an end, however, quite early enough to suit the beaver. They emerge from the pond at the earliest moment that frozen conditions will allow. If their subway is choked with ice, and food becomes exhausted, they will sometimes bore holes through the base of the dam.

Apparently, too, holes of this kind are bored through, or a section cut through the dam to the bottom, for the purpose of completely draining the pond. As this appears to be most often done with ponds that are full of stagnant water, or water almost stagnant, this draining may be a part of the beaver's sanitary work, — done for the purpose of getting filth and stale water out and also that the sour bottom may be sterilized by sun and wind.

Conditions determine the length of time before the dam is repaired and the pond refilled. In some cases this is done after the lapse of a few weeks

The Colony in Winter

and in others not until autumn. Ponds that have large pure streams running through them do not need this emptying, but occasionally they accidentally have it. Most beaver colonies are deserted in summer, and fall thus into temporary decline.

By late summer or early autumn the beaver have assembled at the place where the winter is to be spent. There are patriarchs, youngsters, and those in the prime of life. Around the old home are many who set forth from it when the violets were blooming, when the grass was at its greenest, and when mated birds were building. During the summer a few perished, while others cast their lot with other established colonies. A few of the younger make a start for themselves in new scenes,—found a new colony. Again the dam is repaired and the house recovered; again the harvest home, and again a primitive home-building family are housed in a hut that willing hands have fashioned. Again the pond freezes, and again the snow falls upon a home that stands in a valley where countless generations of beaver have lived through ice-bound winters and the ever-changing happy seasons.

The Original Conservationist

The Original Conservationist

To "work like a beaver" is an almost universal expression for energetic and intelligent persistence, but who realizes the magnitude of the beaver's works? What he has accomplished is not only monumental but useful to man. He was the original Conservationist. An interesting and valuable book could be written concerning the earth as influenced and benefited by the labors of the beaver. The beaver is intimately associated with the natural resources, soil, and water. His work is not yet done, and along the sources of innumerable streams he will ever be needed to save soil, to regulate stream-flow, and to provide pools for the fish.

The beaver's conservation work is accomplished by means of the dams he constructs across streams of flowing water and the ponds that are thus formed. These dams and ponds render a number of services: first, they save soil; second, they check erosion; third, they reduce flood damage; fourth, they store water and help to sustain stream-flow; fifth, they provide water-holes for fish; and sixth, they are helpful in maintaining deep waterways by reducing the extremes of both high and low water, and also by reducing the quantity of sediment carried down into river-channels.

I had enjoyed the ways of "our first engineers" before it dawned upon me that their works might be useful to man, and that the beaver through his constructive handling of the natural resources might justly be called a conservationist. One dry winter the stream through the Moraine Colony ran low and froze to the bottom, and the only trout in it that survived were those in the deep holes of beaver ponds. These ponds offer many advantages to fish multiplication. Much food acceptable to the fish is swept into these ponds. Altogether a beaver pond is an excellent local habitation for fish.

One gray day while I was examining a beaver colony there came another demonstration of the

The Original Conservationist

usefulness of beaver ponds. The easy rain of two days ended in a heavy downpour—a deluge upon the mountain-side a mile or so upstream. There was almost nothing on this mountain either to absorb or delay the excess of water which was speedily shed into the stream below. Flooding down the stream's channel above the beaver pond, came a roaring avalanche of water, or water-slide, with a rubbish-filled front that was five or six feet high. This expanded as it rolled into the pond, and swept far out on the sides, while the water-front, greatly lowered, rushed over the dam. A half a dozen ponds immediately below sufficed so to check the speed of this water and so greatly to reduce its volume that as it poured over the last dam of this colony it was no longer a flood.

The regulation of stream-flow is important. There are only a few rainy days each year, and all the water that flows to the sea through river-channels falls during these few rainy days. The instant the water reaches the earth it is hurried away by gravity, and unless there are factors to delay this run-off, the rivers would naturally contain water only on the rainy days and for a little while thereafter. A beaver dam and pond together form a factor of importance in the keeping of streams ever flowing. The pond is a reservoir which catches and retains some of the water coming into it during rainy days and which delays the water-flow through it. A beaver pond is a leaky reservoir, a kind of spring as it were, and if stored full during rainy days the leakage from it will help maintain stream-flow below during the dry weather. Beaver works thus tend to distribute to streams a moderate quantity of water each day. In other words they spread out or distribute the water of the few rainy days through all the days of the year.

A river which flows steadily throughout the year is of inestimable value to mankind. If floods sweep a river, they do damage. If low water comes, the wheels of steamers and of factories cease to move, and a dry river-channel means both damage and death. Numerous beaver colonies along the sources of countless streams that rise in the hills and the mountains would be helpful in equalizing the flow of these streams. I hope and believe that before many years every rushing

The Original Conservationist

care-free brook that springs from a great watershed will be steadied in a poetic pond that is made, and that will be maintained by our patient, persevering friend the beaver.

In the West beaver are peculiarly useful at stream-sources, where their ponds store flood waters that may later be used for stock water or for irrigation purposes. There are a number of localities in New Mexico, South Dakota, and elsewhere in the West where beaver receive the utmost protection and encouragement from ranchers, whose herds are benefited by water conveniently stored in beaver ponds. A few power companies in the country have commenced to stock with beaver the watersheds which supply them with water. They do this because they realize that countless small ponds or reservoirs are certain to be constructed by these little conservationists.

Running water dissolves and erodes away the earthy materials with which it comes in contact. The presence of a beaver pond and dam across a stream's highway prevents the wearing and the carrying away of material. They not only prevent erosion or wearing away, but they take soil and sediment from the water which comes to them and thus cause an upbuilding. Hence the presence of beaver ponds along streams causes an accumulation of sediment and soil. In time these fill rocky channels and cañons, widen and lengthen valleys, and thus extend the productive area of the earth.

Beaver ponds are settling-basins, and in them are deposited the heavier matter brought in by the stream. In time the pond is filled, and if the beaver do not raise the height of the dam, the accumulated earthy matter becomes covered with flowers or forests.

On the headwaters of the Arkansas River in Colorado some placer miners found gold in the sediment of an inhabited beaver pond. In washing out the deposit of the pond they broke into an enormous amount of loose material beneath, that apparently had been piled in there by glacial action. This material, when removed, was found to have been resting in an ancient beaver pond that was about thirty feet below the one at the surface.

The Original Conservationist

A few centuries ago there were millions of beaver ponds in North America; most of these were long since filled with sediment. Since then, too, countless others have been formed and filled. This soil-saving and soil-spreading still goes ever on wherever there is a beaver pond.

Many of the richest tillable lands of New England were formed by the artificial works of the beaver. There are hundreds of valleys in Kansas, Kentucky, Missouri, Illinois, and other States whose rich surface was spread upon them by the activities of beaver through generations. In the Southern States and in the mountains of the West, the numbers of beaver meadows are beyond computation. The aggregate area of rich soil-deposits in the United States for which we are indebted to the beaver is beyond belief, and probably amounts to millions of acres.

The beaver have thus prepared the way for forests and meadows, orchards and grain-fields, homes and school-houses. In the golden age of the beaver, their countless colonies clustered all over our land. These primeval folk then gathered their harvest. Innumerable beaver ponds, which then shone everywhere in the sun, slowly filled with deposited, outspreading soil,—and vanished. Elm avenues now arch where the low-growing willow drooped across the canal, and a populous village stands upon the seat of a primitive and forgotten colony.

A live beaver is more valuable to mankind than a dead one. As trappers in all sections of the country occasionally catch a beaver, it is probable that there still are straggling ones scattered along streams all the way from salt water up to timber-line, twelve thousand feet above sea-level. These remaining beaver may be exterminated; but if protected they would multiply and colonize stream-sources. Here they would practise conservation. Their presence would reduce river and harbor appropriations and make rivers more manageable, useful, and attractive. It would pay us to keep beaver colonies in the heights. Beaver would help keep America beautiful. A beaver colony in the wilds gives a touch of romance and a rare charm to the outdoors. The works of the beaver have ever intensely interested the human mind. Beaver works may do for

The Original Conservationist

children what schools, sermons, companions, and even home sometimes fail to do, — develop the power to think. No boy or girl can become intimately acquainted with the ways and works of these primitive folk without having the eyes of observation opened, and acquiring a permanent interest in the wide world in which we live. A race which can produce mothers and fathers as noble as those beaver in the Grand Cañon who offered their lives hoping thereby to save their children is needed on this earth. The beaver is the Abou-ben-Adhem of the wild. May his tribe increase!

THE END

Bibliographical Note

BEAVER literature is scarce. The book which easily excels is "The American Beaver and his Works," by Lewis H. Morgan. Samuel Hearne has an excellent paper concerning the beaver in "Journey from Prince of Wales Fort to the Northern Ocean," published in 1795. Good accounts of the beaver are given in the following books: "Beavers: their Ways," by Joseph Henry Taylor; "Castorologia," by Horace T. Martin; "Shaggycoat," by Clarence Hawkes; "The House in the Water," by Charles G. D. Roberts; and "Forest Neighbors," by William Davenport Hulbert. There are also admirable papers by Ernest Thompson Seton in his "Life-Histories of Northern Animals," by W. T. Hornaday in his "American Natural History," and by Baillie-Grohman in "Camps in the Rockies."

Notes

Notes are keyed to page and line numbers. For example, 4:3 means page 4, line 3.

Frontispiece. The photographs and illustrations in this book, unless otherwise indicated, are those of Enos Mills.

Dedication: J. Horace McFarland (1859–1948) was the president of the powerful American Civic Association, and at the time of the book's publication Mills's ally in the campaign that would succeed two years later in creating the Rocky Mountain National Park. Their relationship would later dissolve in controversy over the vehemency of Mills's attack on the National Park Service. See the Papers of J. Horace McFarland, File 80, Division of Public Records, Pennsylvania Historical Museum Commission, Harrisburg, Pennsylvania.

"Working Like a Beaver"

11:8. The Jefferson River in southwestern Montana rises in the Granelly Range near Yellowstone National Park.

"Our Friend the Beaver"

22:4–6. The Lewis and Clark trail in western Montana passed close to the future site of Butte where, as the introduction notes, Enos Mills worked for a number of years as a seasonal miner.
25:3. The Snake River, 1,038 miles in length and the chief tributary of the Columbia, rises in northwest Wyoming in Yellowstone National Park.
28:10. The Moraine Colony, located some three and a half miles from Enos Mills's Longs Peak Inn, consisted of some fifty ponds of varying size. The colony was fed by the Roaring Fork, a stream that flows out of Chasm Lake, at the base of Longs Peak, between two moraines to Cabin Creek in the valley below. The beaver dams of the Moraine Colony began some three miles below Chasm Lake and extended for about three quarters of a mile, at an elevation of 8,850 to 9,000 feet.

33:13–14. The reference to beavers "at work in broad daylight" is found in Edwin James's *Account of an Expedition from Pittsburgh to the Rocky Mountains . . . Under the Command of Stephen H. Long* (Philadelphia: Carey and Lea, 1823), I, 464n. As indicated, the Long expedition took place in 1819–20.

"The Beaver Past and Present"

40:16–17. Mills visited the Seine during his summer 1900 trip to the Paris Exposition.
41:1–10. Mills unquestionably gleaned these facts from Horace T. Martin's *Castorologia: Or the History and Traditions of the Canadian Beaver* (Montreal, 1892), pp. 5, 28. It was an important early work on the North American beaver, which Mills quotes from below on page 49.
42:7–10. See Chapter XIII, *History of the Expedition Under the Command of Captains Lewis and Clark, to the Sources of the Missouri, Thence Across the Rocky Mountains* (Philadelphia: Bradford and Inskeep, 1814).
42:13. 1885 was Enos Mills's second year in Estes Park.
48:12–22. Mills is indebted here to Clarence Hawkes's *Shaggycoat: The Biography of a Beaver* (Philadelphia: George W. Jacobs and Company, 1906), p. 17.
48:23. William Kingsford (1819–1898), *The History of Canada*, 10 volumes (1887–98).
49:3–4. As noted above, Horace T. Martin was the author of *Castorologia: Or the History and Traditions of the Canadian Beaver* (Montreal: William Drysdale and Company, 1892). The quotation that Mills attributes to Martin is found on the page preceding the title page, where it is followed by the name "John Reade."
49:11–12. The American Fur Company was incorporated in 1808 to bring together John Jacob Astor's widespread fur trading interests in the West. By 1817 he had gained control of the Northwest Company as well. Astor (1763–1848) used his virtual monopoly of the beaver trade to amass the largest private fortune in America.

"As Others See Him"

53:15–17. "We are now almost led to regret that three-fourths of the old accounts of this extraordinary animal are fabulous. . . ." John James Audubon (1785–1851), *The Quadrupeds of North America* (New York: V. G. Audubon, 1854), I, 349.
53:20. Samuel Hearne, *A Journey from Prince of Wales's Fort in Hudson's Bay* (London: Cadell and Davies, 1795), p. 231. Mills undoubtedly came across the quotation from Hearne during his reading of either Lewis H.

Morgan's *The American Beaver and His Works* (Philadelphia: J. B. Lippincott, 1868) or Horace T. Martin's *Castorologia* (1892), previously cited above. Both books include the excerpt from Hearne's Journey that includes the quotation in question as an appendix. Morgan (1818–1881), a lawyer in Rochester, New York, became interested in the beaver while serving as legal adviser to a railroad running from Marquette, Michigan, to the iron region on the southern shore of Lake Superior. His book of 1868, like Horace Martin's, has long been regarded as a classic on the subject.

54:21–55:9. Morgan, p. 18.
55:15–56:2. Morgan, p. 9.
57:14–17. Enos A. Mills, *Wild Life on the Rockies* (Boston: Houghton Mifflin Company, 1909), p. 59.
58: 8–10. Morgan, p. 250.
58:11–13. Morgan, p. 262.
58:13–18. The precise source of this quotation is unclear, though Morgan does conclude on page 264 that "These several artificial works show a capacity in the beaver to adapt his constructions to the particular conditions in which he finds himself placed. Whether or not they evince progress in knowledge, they at least show that the beaver follows, in these respects, the suggestion of a free intelligence."
58:20–59:17. George J. Romanes (1848–1894), *Animal Intelligence* (New York: D. Appleton and Company, 1906), pp. 367, 376. Romanes indicates his own indebtedness to Morgan's *The American Beaver* "for my statement of facts" (367).
59:21–60:18. Alexander Majors (1814–1900), *Seventy Years on the Frontier: Alexander Majors' Memoirs of a Lifetime on the Border* (Chicago: Rand, McNally, 1893), p. 218. Majors devotes an entire chapter of his narrative, pp. 215–20, to "The Beaver."
60:23–61:27. Henry Wadsworth Longfellow (1807–1882), "The Song of Hiawatha" (1855), Introduction, 1–4, 22–27; XVII, 43–59.

"The Beaver Dam"

80:1. At Three Forks in southwestern Montana the Gallatin, Jefferson, and Madison rivers come together to form the Missouri. The future site of the town of Three Forks was visited by Lewis and Clark.

"Harvest Time with Beavers"

83:4. The Spruce Tree Colony, part of the Moraine Colony, was probably so named because of its proximity to several stands of Engelmann spruce that were here interspersed with the more customary lodgepole pines, willows, alders and aspens. Edward R. Warren, who visited Spruce Tree Colony during the summer of 1922, described it as follows: "This pond has a dam which I estimated to be at least 300 feet long, and contains a fine lodge, which was inaccessible, being out in the water away from the shore. After careful study from a point on the dam directly opposite the lodge I estimated its north and south diameter to be at least twenty feet on the waterline, and the height above water 5 feet. . . ." Warren, p. 210.

83:6–7. Now known as Mills Moraine after the naturalist-author.

92:24–93:1. Island Colony, situated in a pond about 100 by 200 feet, was located on the North Fork of Cabin Creek, which rises on the southeasterly flank of Longs Peak.

"Transportation Facilities"

101:2. Lily Lake, two miles to the north of Longs Peak Inn, lies at the foot of Lily Mountain (9,786 feet) on the road to Estes Park village at an elevation of 9,000 feet. The lake drains into the Big Thompson River by way of Fish Creek.

102:20. Wind River is a small stream in the Glacier Basin region of Rocky Mountain National Park.

"The Primitive House"

128:11–12. Longs Peak (14,256 feet) and its near neighbor to the south Mount Meeker (13,911 feet) are the two highest mountains in Rocky Mountain National Park.

"The Beaver's Engineering"

140: 11–12. Mills may mean 1901. "Early in the summer of 1901 a man in burning some logs out of a trail on the eastern slope of Longs Peak neglected the fire and it spread and killed more than a thousand acres of beautiful, valuable forest." Enos A. Mills, *The Story of Estes Park and A Guide Book* (Denver: Outdoor Life Publishing Co., 1905), p. 53.

140: 17–18. Bierstadt Moraine, located in the Glacier Basin region of Rocky Mountain National Park near Bear Lake, was named after the German-born landscape artist Albert Bierstadt (1830–1902), whom the Earl of Dunraven (see note 179:8, below) brought to Estes Park in 1876 and 1877 to paint on commission Longs Peak. The result is the stunning canvas entitled "Rocky Mountains, Longs Peak," which once adorned the walls of Dunraven Castle in Ireland and now hangs in the Western Room of the Denver Public Library.

142:19. Chasm Lake lies in a glacial cirque at the base of the sheer East Face of Longs Peak at an elevation of 11,760 feet.

150:17–18. By the summer of 1922, however, "dead wood dam," as Mills called it, had come back to life. See Warren, p. 209.

"Beaver Pioneers"

176:24. Presumably a stream in Willow Park, as Moraine Park was originally called.

179:8. Windham Thomas Wyndham-Quin (1841–1926), the fourth Earl of Dunraven, a wealthy lord, visited Estes Park on a hunting expedition in late December 1872 and again in 1873 and 1874. He later launched an ambitious scheme to gain control of the entire valley for his exclusive use as a game preserve. Though Lord Dunraven's grand design was foiled by the militancy of early settlers, he did succeed in gaining control of upwards of 15,000 acres, and built the three-story fifty-room English (or Estes Park) Hotel on lower Fish Creek Road as well as a private lodge nearby.

"The Colony in Winter"

197:1. The Medicine Bow Mountains of Colorado and Wyoming lie northwest of Estes Park.

"The Original Conservationist"

218:15–16. The Arkansas River rises on the east slope of the Rockies in central Colorado near Leadville.

221: 12–13. "Abou Ben Adhem (may his tribe increase!)" is the first line of a popular poem written in 1838 by English poet Leigh Hunt (1784–1859).

222: 2–3. Lewis H. Morgan (1818–1881), *The American Beaver and His Works* (Philadelphia: J. B. Lippincott and Company, 1868).

222: 3–6. Samuel Hearne (1745–1792), *A Journey from Prince of Wales's Fort in Hudson's Bay to the Northern Ocean* (London: Cadell and Davies, 1795). The "excellent paper concerning the beaver" occurs at the end of Chapter VII.

222: 8–9. Joseph Henry Taylor (1845–?), *Beavers: Their Ways and Other Sketches* (Washburn, N.D.: Joseph H. Taylor, 1904).

222: 9. Horace T. Martin (1850–c.1905), *Castorologia, or The History and Traditions of the Canadian Beaver* (Montreal: Drysdale; and London: E. Sanford, 1892).

222: 10: Clarence Hawkes (1869–1954), *Shaggycoat: The Biography of a Beaver* (Philadelphia: G. W. Jacobs and Company, 1906). Hawkes uses as his preface the same lines from Longfellow's Hiawatha that Mills quotes on page 61.

222: 10–11. Charles G. D. Roberts (1860–1943), *The House in the Water: A Book of Animal Stories* (Boston: L. C. Page and Company, 1908).

222: 12–13. William Davenport Hulbert (1868–1913), *Forest Neighbors; Life Stories of Wild Animals* (New York: McLure, Phillips and Company, 1902). See "The Biography of a Beaver," pp. 1–40.

222: 13–15. Ernest Thompson Seton (1860–1946), *Life-Histories of Northern Animals: An Account of the Mammals of Manitoba* (New York, Charles Scribner's Sons, 1909). See Volume I, Chapter XVI, "Canadian Beaver," pp. 447–79.

222: 15–16. William Temple Hornaday (1854–1937), *The American Natural History: A Foundation of Useful Knowledge of the Higher Animals of North America* (New York: Charles Scribner's Sons, 1904).

222: 16–17. William A. Baillie-Grohman (1851–1921), *Camps in the Rockies. Being a Narrative of Life on the Frontier, and Sport in the Rocky Mountains* (New York: Scribner's, 1882). See Chapter IX, "The Beaver and His Camp."

Index

Accidents, 144.
Age, 14, 193.
Air, blanket over pond, 202, 203.
American Fur Company, 49.
Arkansas River, 218.
Astor, John Jacob, 49.
Attitudes, 6.
Audubon, John James, 53.
Autumn activities, beginning of, 200.

Bad Lands, 65.
Basins, food, 108. *See also* Wells.
Beaver, a tame, 22-25.
Beaver, aged, of the Spruce Tree Colony, 83, 84, 95, 96; of Lily Lake, 102-105; migrating to the Moraine Colony, 167, 168.
Bedding, 122, 123.
Bierstadt Moraine, 140.
Bobcat, 35.
Burrows, 110, 111; a substitute for houses, 127, 128.

Canada, emblem of, 43.
Canals, 77, 78, 88, 141, 145-149, 187; at Lily Lake, 103, 104; importance, 105; use of excavated material, 105, 106; forms of, 106, 107; system at Three Forks, Mont., 107-111; dug in winter, 206.
Castoreum, 43, 44.
Chasm Lake, 142.
Civilization, the beaver's influence on, 47-49.
Color, 8.

Colorado River, 25, 50.
Coöperation, 171.
Coyotes, 23, 102, 161-163, 166.
Cry, 27.
Cutting trees, methods of, 10-12, 31, 32; intelligence shown in, 57, 91; operations observed, 86, 90-96; accidents in, 144.

Dams, materials, 65-67; construction, 66, 67; uses, 69; growth, 69, 70; new and old, 70, 71; discharge from, 71, 72; not all beaver build, 72; thoroughfares, 73; effect on topography, 73, 74; shape, 75-77; an interesting dam, 76-78; waterproofing, 78; dimensions of a long dam, 78, 79; dimensions of other dams, 86; across canals, 108-110; the dead-wood dam, 143-150; across a drainage ditch, 180, 181; across an irrigation ditch, 189, 190; a homesteader's dam completed by beaver, 192, 193; effect on stream-flow, 213-217.
Day, working by, 33, 94, 156.
Death, 14.
Ditch, struggle over a, 179-182.
Ditches. *See* Canals.
Diver, the young beaver, 22-25.
Domestication, 25.
Dunraven, Lord, 179.

Ears, 7.
Enemies, 14; times of danger from, 198.

Index

Engineering, 139–150.
Erosion, checked by beaver, 214, 217, 218.
Errors, 67, 68.
Estes Park, 179.
Europe, the beaver in, 40, 41.
Exploration, 168, 169.
Eyesight, 8.

Fabulous accounts, 53.
Feet, uses of, 5, 6; form of, 8.
Feigning injury, 25, 26.
Felling trees. *See* Cutting trees.
Fence posts, 30.
Fighting, 19, 20, 34, 35.
Fire, 158–163.
Fish, water-holes for, 214.
Flat-top, a beaver pioneer, 175, 176, 183–193.
Floods, 206, 207; damage prevented by beaver, 214, 216.
Food, 10, 84, 205.
Food-piles, 12, 13, 88, 89, 97, 150, 169.
Fossil beaver, 40.
Fox, 199.
Fruit trees, 30.

Geographical distribution, 40–42, 49, 50.
Gold, 218.
Grand Cañon, 25, 50.

Hands, uses of, 5; form of, 8.
Harvest, a year's, 97; a large, 169.
Harvest-gathering, 83–98, 148–150, 157, 158.
Hearing, 8.
Hearne, Samuel, quoted, 53.
History, the beaver in, 41–44.
Homesteader, a friendly, 190–193.
Houses, building, 3; occupants, 21; dimensions, 86, 119, 120, 130, 131; mud plastering, 97, 123–125; construction, 119–123, 130, 131; entrances, 119, 120; situation, 120, 125–127; burrows a substitute for, 127, 128; a typical house, 130, 131; ventilation, 132; enlargement, 169–171; security, 197, 198; shaped to meet floods, 207.
Hudson's Bay Company, the, 48.

Ice, a trouble of beaver existence, 126, 127; a catastrophe caused by, 184–186; on the pond, 200, 202–206; casualties caused by, 207.
Indians, their legends about the beaver, 39.
Individuality, 35, 67.
Industry, 36.
Intelligence, 46, 57–60.
Irrigation-ditches, 31.
Island Colony, harvesting methods of, 92, 93.

Jefferson River, 11, 78, 107, 108.

Kingsford, William, his History of Canada, 48.

Land, beaver seen on, 191, 192.
Leadership, 20.
Legends, 39.
Lewis and Clark, 42.
Life, the beaver's, 14–16.
Lily Lake, beaver at, 101–105; beaver house at, 119; the pioneer beaver of, 175–193; description of, 178.
Lily Mountain, 182.
Lion, mountain, 160–162, 166.
Local attachment, 141, 142.
Long, Stephen Harriman, his Journal, 33.
Longfellow, Henry Wadsworth, his *Hiawatha* quoted, 60, 61.
Long's Peak, 140, 153.
Love ditty, 27.

Index

Majors, Alexander, his *Seventy Years on the Frontier*, 59, 60.
Martin, Horace T., 49.
Mating, 27.
Medicine Bow Mountains, 197.
Migration, 20, 21, 132, 133, 141, 161-163, 167-169, 175-177, 182, 183.
Mischief, 30, 31.
Moraine Colony, engineering of, 139, 142-150; discovery and observation of, 153-158; homes destroyed by fire, 158, 159; migrating, 161-163; new site, 163, 164; old site resettled, 165; later fortunes, 166-171.
Morgan, Lewis H., his *American Beaver and his Works*, 54, 55, 58.

Night, working at, 33.
Northwestern Fur Company, 49.

Old, the, 34.
Outcasts, 34.
Ouzel, water, 199.

Parasites, 14.
Physical make-up, 5-9, 68.
Pipestone Creek, 11.
Place-names taken from the beaver, 42, 43.
Play, 29, 156, 157.
Ponds, early abundance, 42; size, 65, 86; uses, 68, 69; chains or clusters of, 74; depth, 107; canals in bottom, 107; spring-filled, 113, 114; lowering the level under ice, 202, 203; draining, 208, 209; effect on stream-flow, 213-217; leaky reservoirs, 216.
Population, changes in, 46, 47.
Protection, 50, 217, 220, 221.

Reason, evidences of, 57, 58.
Romanes, George J., on the beaver, 58, 59.

Sanitation, 208.
Sawtooth Mountains, 66.
Sediment, a problem of beaver life, 125, 126.
Sheep, mountain, 192.
Size, 7.
Skins, 43, 44, 48, 49.
Sleep, 122.
Slides, 87, 112, 199.
Smell, sense of, 7.
Snake River, 25.
Soil, the beaver's conservation of, 214, 217-220.
Sounds and silence, 19, 20, 23, 26, 27, 133, 134.
Springs, use of, 204.
Spruce Tree Colony, harvest time with, 83-98; tunnels in, 113-115.
Stream-flow, effect of beaver on, 72-74, 213-217.
Strength, 9.
Subways. *See* Tunnels.
Swimming, method of, 6.

Tail, uses of, 5, 6, 11; form and covering, 8; signalling with, 24, 31, 96; fabulous accounts of the uses of, 53.
Teeth, 7-9.
Three Forks, Montana, 42, 79; canal system at, 107-111.
Trails, 111, 112.
Transportation of dam and food material, 86-90, 92, 93; canals used in, 106-115; trails and slides used in, 111, 112, 115; tunnels used in, 112-115.
Trappers, 164, 189-191.
Traps, 35, 189.
Trees, cutting. *See* Cutting trees.
Trimming trees, 12, 96.
Trout, 205, 206.
Tunnels, 85, 112-115, 198, 203, 206.

233

Index

Water. *See* Stream-flow.
Water-ouzel, 199.
Water-supply, 85, 86.
Weather-wisdom, 44–47.
Weight, 7.
Wells, food, 103, 104. *See also* Basins.
Whistle, 26, 27.
Wildcat, 35.

Willow Creek, 176.
Wind River, 102, 175, 182, 188.
Winter, beaver life in, 197–209.
Wolves, gray, 191, 197.
Wood, dead, 143, 144.
Work, accomplished by beaver, 3–5.

Young, birth and care of, 27, 28; growth and play of, 28.